RAINWATER HARVESTING

RAINWATER HARVESTING

Dr. Chandrawati Jee

Shagufta

A P H PUBLISHING CORPORATION
4435-36/7, ANSARI ROAD, DARYA GANJ
NEW DELHI-110 002

Published by

S.B. Nangia
APH Publishing Corporation
4435-36/7, Ansari Road, Darya Ganj
New Delhi-110 002
PH. : 23274050
Email : aphbooks@vsnl.net

2013

Typesetting at

Rawat Computers
Gandhi Nagar, Delhi

Printed at

Balaji Offset
Navin Shahdara, Delhi-32

Preface

Water is one of the most important natural resource, it is the most basic need for all living beings as well as a very valuable national asset. In the planet Earth, water availability is on decline with most obvious reasons as increasing demand for water with increasing rate of population. Moreover the rivers are getting dried up, groundwater is depleting and canals are polluted. The increased demand has not been matched by efforts to conserve and better manage water supplies, meaning that the resource is being depleted faster than it can be replenished.

Considering the present circumstances, the world faces a bleak water future, and might face a severe water crisis within the next two decades, unless water management practices are changed or improved on priority basis. The Earth is blessed with adequate rainfall as a whole, yet there are large swathes of dry, drought prone area. In many other places the quality of groundwater is not good. In such places rainwater harvesting can provide lifeline water for survival and more.

Rainwater harvesting as a method of utilising rainwater for domestic and agricultural use is already widely used throughout the world. It is a method which has been used since ancient times and is increasingly being accepted as a practical method of providing potable water in development projects throughout the world. Rainwater harvesting is regaining its importance as governments are exercising special efforts to promote the same supported by the efforts of NGOs, who trying their best to create awareness in society to understand the significance of rainwater harvesting.

The present volume is a practical guide for rainwater harvesting. It provides authentic information on rainwater harvesting strategies; including site planning with entire chapters on berms, basins, terraces, mulching and much more. It is packed with real life examples, case studies and illustrations. This book will serve as a valuable guide for students and professional of environmental studies, government agencies and NGOs working for water management and anyone serious about living a more sustainable lifestyle.

Author

Contents

1 Introduction

Rainwater harvesting is the gathering, or accumulating and storing, of rainwater. It has been used to provide drinking water, water for livestock, water for irrigation or to refill aquifers in a process called groundwater recharge. Rainwater collected from the roofs of houses, tents and local institutions, or from specially prepared areas of ground, can make an important contribution to drinking water. In some cases, rainwater may be the only available, or economical, water source. Rainwater systems are simple to construct from inexpensive local materials, and are potentially successful in most habitable locations. Roof rainwater is usually of good quality and does not require treatment before consumption. Household rainfall catchment systems are appropriate in areas with an average rainfall greater than 200mm per year, and no other accessible water sources.

Rainwater harvesting is an ancient technique enjoying a revival in popularity due to the inherent quality of rainwater and interest in reducing consumption of treated water. Rainwater is valued for its purity and softness. It has a nearly neutral pH, and is free from disinfection by-products, salts, minerals, and other natural and man-made contaminants. Plants thrive under irrigation with stored rainwater. Appliances last longer when free from the corrosive or scale effects of hard water. Users with potable systems prefer the superior taste and cleansing properties of rainwater.Archeological evidence attests to the capture of rainwater as far back as 4,000 years ago, and the concept of rainwater harvesting in China may date back 6,000 years. Ruins of cisterns built as early as 2000 B.C. for storing runoff from hillsides for agricultural and domestic purposes are still standing in Israel.

1.1 Importance of Rainwater Harvesting

Water is a precious natural resource that needs to be managed and used wisely. We need to do so because the total amount of rainfall in India remains the same, while the increase in population and the growing economic prosperity has created a huge demand for water which has increased the pressure on its availability. Agriculture continues to remain the single largest consumer of water and industrial demand is also growing at a rapid pace.

In urban areas, the demand for water is more concentrated which means that Urban Local Bodies have to pump it from sources that are located at greater distances from the city limits. Many households also depend on groundwater to meet their requirements. There are also many problems being reported with the groundwater quality. For example, the presence of nitrates, fluoride, iron and salts are on the rise.Paradoxically, when it rains many parts of the city gets flooded. This is because rainwater is not managed carefully. Increasing paved areas increases runoff and reduces percolation and the recharge of ground water. Valleys and tanks are being encroached and the smooth flow of rainwater is being disturbed.Hence, incorporating rainwater harvesting techniques while planning cities will not only prevent urban flooding but also recharge groundwater.

Depletion of groundwater and flooding during the monsoon are signals that tell us that urban areas have to be careful with their water use. Water can no longer be wasted and citizens should participate in the process of finding a solution to the problem of water scarcity and contribute their mite to help.

The first thing that you need to understand about water is that the primary source of water is rain. Water evaporating from both the sea and land, falls back as precipitation and runs off as streams and rivers, settles in lakes and tanks, seeps through the soil and reaches the underground aquifer. This water then becomes available to be taken out as well water or bore-well water, but in all cases, the primary source of this water is rain. Rainfall is measured in millimeters using specified rain gauges set up throughout the country. The rainfall for a particular place is averaged over 30 years and reported as normal rainfall for a place.

Based on the actual average rainfall for a city, the theoretical model to be followed for a city level rainwater harvesting system would involve:

- The capturing of rooftop rainwater
- Collecting water runoff from open areas
- Collecting the city level runoff

There are many reasons why rainwater needs to be harvested. Some of them are given below:

- Water is a scarce natural resource and hence it should not be wasted.
- It can supplement our domestic, industrial and other water needs.
- If rainwater harvesting is incorporated at the design stage of any construction project, the investment is low and the water collection is large.
- Centralised systems of water supply are under stress to cope with huge water supply demand of cities.
- Ground water is constantly getting depleted or polluted in large areas.
- Water is not only becoming scarce but expensive as well.
- The link with the hydrologic cycle is re-established leading to efficient and effective use of water.
- It can be an individual's contribution to reduce the 'ecological water footprint' i.e. the distance water has to travel to reach a house.

- It will reduce electricity consumption and the cost required to pump water from sources that are a long distance away.
- It will mitigate and to a certain extent prevent urban flooding.

Consider the example of a house being built on a new housing site. The site would have been naturally vegetated land or agricultural land prior to the construction. The local hydrologic cycle would change dramatically after the construction.

One of the aims of rainwater harvesting is to restore the original hydrologic cycle by storing the increased runoff and by using some of it for recharging.

The restoration of the natural hydrologic cycle is the key aim of rainwater harvesting. This is true even for small housing sites, as it is, for the entire city.

Change in water flows with increasing paved surface:

- Infiltration is the process through which rainwater enters the soil.
- Percolation is the process through which it proceeds to the aquifer below.
- Evaporation is the process by which rainwater goes back to the atmosphere as water vapour.
- Transpiration is the process through which plants breathe out water vapour.
- Evapotranspiration is the combination of evaporation and transpiration.

Rainwater harvesting can have manifold reasons. Some of the reasons rainwater harvesting can be adopted in cities are to provide supplemental water for the city's requirements, to increase soil moisture levels for urban greenery, to increase the ground water table through artificial recharge, to mitigate urban flooding and to improve the quality of groundwater. In urban areas of the developed world, at a household level, harvested rainwater can be used for flushing toilets and washing laundry. Indeed in hard water areas it is superior to mains water for this. It can also be used for showering or bathing. It may require treatment prior to use for drinking.

1.2.1. Water Consumption and Role of Rainwater

It is a good idea to install a water meter and measure monthly consumption of water, whether it is from the city pipelines or from your own well or bore-well. A good water meter should be Bureau of Indian Standards (BIS) certified and bear the BIS certification mark. Measuring consumption is the first step to managing and conserving water usage. However, the consumption pattern as suggested by the Central Public Health and Environmental Engineering Organisation (CPHEEO) standards is given below:

Table 1.Consumption Pattern

Type of Use	*Litres per person and per day*
Drinking	3
Cooking	4
Bathing	20
Flushing	40
Washing clothes	25
Washing utensils	20
Gardening	23

If you examine the water bill from your service provider, it should give you a fair idea of your water use. Consumption can range from 50 litres to 300 litres per person per day. At many times, you will find that cheap and subsidised water results in inefficient use and high consumption.

How much Water can a Person get from the Rain? It would depend upon where you are located and how much rain falls there.

Harvested Rainwater can be used for:

- Non-potable purposes like gardening, flushing and washing clothes.
- All purposes, but you can do this only after confirming the potable qualities of harvested rainwater.
- Only potable purposes after ensuring the quality of the water, where there is no alternative source or the source is contaminated. This is typical in many habitations.

As the nation is becoming increasingly aware, rainwater harvesting is a building system that collects and conveys rainwater from roofs, filters debris from the water and stores the rainwater for use. Systems can be non-potable or potable, the latter requiring additional filtering after the water is pumped from the tank. Across India, people realise the need to conserve water.

Rainwater harvesting enjoys widespread use in areas around the world such as Europe and the Caribbean islands, and these systems are a logical solution for our country's water resource challenges. There is growing enthusiasm about green, sustainable living systems and practices that save water, energy and money, while preserving and improving our natural environment. Rainwater harvesting supplies alternative water for such non-potable uses as irrigation, toilet flushing, laundry, vehicle and facility cleaning, fire suppression systems, HVAC cooling towers and agriculture. Potable rainwater harvesting systems can be used for drinking water, as well as showers/baths, dishwashing, swimming pools and food processing. Rainwater harvesting is a powerful solution to a range of water problems.

1.2. Process of Harvesting Water

Collecting and storing of rainwater for productive use in the future is called rainwater harvesting.

There are certain other terms which need to be understood when you speak of rainwater harvesting, such as

- rooftop rainwater harvesting which is collection of rainwater from rooftops.
- storm-water harvesting which is collection of rain from paved and unpaved areas and also from storm-water drains.

Water can be collected either from rooftops or from the ground or a combination of both. It may involve the simple collection of rainwater from the rooftop of a house, for use in gardening or collection from a large industrial campus for reuse in industrial processes or for recharging into the ground as shown.

Rainwater can be harvested by anybody.

- It is ideal for people planning to construct or in the process of constructing any building. It could be a home, a school, a factory or a shed.
- Already built structures can be retrofitted or refitted for rainwater harvesting.
- Industries and institutions with large roof areas have a good potential for rainwater harvesting.
- Apartments and multistoried buildings too, can harvest rainwater.
- Paved and unpaved ground also provides a potential catchment for rainwater harvesting.
- In rural areas, where water needs to be brought from a source that is at a long distance away or where the groundwater is contaminated with fluoride, nitrates, arsenic etc.
- In low income areas, where water is to be stored in small containers and is available infrequently.
- Parks and open spaces can harvest rain to increase soil moisture and groundwater levels for meeting their water requirements.
- Cities and towns can harvest rainwater in lakes and tanks to supplement their water requirement.

1.3. Rainwater Harvesting Systems

There are a number of types of systems to harvest rainwater ranging from very simple to the complex industrial systems. Generally, rainwater is either harvested from the ground or from a roof. The rate at which water can be collected from either system is dependent on the plan area of the system, its efficiency, and the intensity of rainfall.

1.3.1. Ground Catchment Systems

Ground catchments systems channel water from a prepared catchment area into storage. Generally they are only considered in areas where rainwater is very scarce and other sources of water are not available. They are more suited to small communities than individual families. If properly designed, ground catchments can collect large quantities of rainwater.

1.3.2. Roof Catchment Systems

Roof catchment systems channel rainwater that falls onto a roof into storage via a system of gutters and pipes. The first flush of rainwater after a dry season should be allowed to run to waste as it will be contaminated with dust, bird droppings etc. Roof gutters should have sufficient incline to avoid standing water. They must be strong enough, and large enough to carry peak flows. Storage tanks should be covered to prevent mosquito breeding and to reduce evaporation losses, contamination and algal growth. Rainwater harvesting systems require regular maintenance and cleaning to keep the system hygienic and in good working order.

1.3.3. Subsurface Dyke

A subsurface dyke is built in an aquifer to obstruct the natural flow of groundwater, thereby raising the groundwater level and increasing the amount of water stored in the aquifer.The subsurface dyke at Krishi Vigyan Kendra, Kannur under Kerala Agricultural University with the support of ICAR, has become an effective method for ground water conservation by means of rainwater harvesting technologies. The sub-surface dyke has demonstrated that it is a feasible method for conserving and exploiting the groundwater resources of the Kerala state of India. The dyke is now the largest rainwater harvesting system in that region.

1.3.4. Groundwater Recharge

Rainwater may also be used for groundwater recharge, where the runoff on the ground is collected and allowed to be absorbed, adding to the groundwater. In the US, rooftop rainwater is collected and stored in sump. In India this includes Bawdis and johads, or ponds which collect the run-off from small streams in wide area. In India, reservoirs called tankas were used to store water; typically they were shallow with mud walls. Ancient tankas still exist in some places.

1.4. Analysis of Rainwater Quality

Many questions arise in people's minds when the issue of rainwater quality comes up. Is it potable? Is the water the result of acid rain? Will it 'spoil' if it is kept in storage for a few days? Will it contain worms and other organisms? Will it be too soft without minerals? Will you get a cold if you drink rainwater?

Drinking water has to meet requirements of the Indian Standards' 10500 for physical, chemical and biological parameters. Certified government and private laboratories can check the water sample to ensure that it is potable. In Karnataka, the Department of Mines and Geology and the Karnataka State Pollution Control Board provide water testing facilities in many towns. In all cases, check rainwater to make sure that it is potable before drinking.

Remember that for most of the time you are not going to store water for use in summer. What we suggest is that when it rains, harvest rainwater and use it first. If it replaces groundwater, you have saved groundwater for a rainy (sunny) day. If it replaces piped water, energy cost in pumping the water has been saved. However, if indeed, you are storing water for a sunny day, here is what could happen.

Water needs three conditions to go bad: air, sunlight and organic matter within it.

1. By excluding organic matter through proper filtration and by storing water in a closed container without access to sunlight and air, water will remain good for a long time.
2. Treating water with a solution of 30 gms of alum, dissolved in 3 liters of water and added to 250 liters reduces turbidity of water.
3. Bleaching powder: Make a solution by adding one large spoon in two litres of water and this solution has to be added to about 1000 litres of water as an effective treatment for bacterial contamination.

Water should always be treated well before it is drunk and wherever possible, boiled or sterilised.

As rainwater may be contaminated, it is often not considered suitable for drinking without treatment. However, there are many examples of rainwater being used for all purposes – including drinking – following suitable treatment. Rainwater harvested from roofs can contain animal and bird faeces, mosses and lichens, windblown dust, particulates from urban pollution, pesticides, and inorganic ions from the sea (Ca, Mg, Na, K, Cl, SO_4), and dissolved gases (CO_2, NOx, SOx). High levels of pesticide have been found in rainwater in Europe with the highest concentrations occurring in the first rain immediately after a dry spell;the concentration of these and other contaminants are reduced significantly by diverting the initial flow of water to waste as described above. The water may need to be analysed properly, and used in a way appropriate to its safety. In the Gansu province for example, harvested rainwater is boiled in parabolic solar cookers before being used for drinking.

1.5. Benefits of Rainwater Harvesting

Advantages and benefits of rainwater harvesting are numerous.

- The water is free; the only cost is for collection and use.
- The end use of harvested water is located close to the source, eliminating the need for complex and costly distribution systems.
- Rainwater provides a water source when groundwater is unacceptable or unavailable, or it can augment limited groundwater supplies.
- The zero hardness of rainwater helps prevent scale on appliances, extending their use; rainwater eliminates the need for a water softener and the salts added during the softening process.
- Rainwater is sodium-free, important for persons on low-sodium diets.
- Rainwater is superior for landscape irrigation.
- Rainwater harvesting reduces flow to stormwater drains and also reduces non-point source pollution.
- Rainwater harvesting helps utilities reduce the summer demand peak and delay expansion of existing water treatment plants.
- Rainwater harvesting reduces consumers' utility bills.

Perhaps one of the most interesting aspects of rainwater harvesting is learning about the methods of capture, storage, and use of this natural resource at the place it occurs. This natural synergy excludes at least a portion of water use from the water distribution infrastructure: the centralised treatment facility, storage structures, pumps, mains, and laterals.

Rainwater harvesting also includes landbased systems with man-made landscape features to channel and concentrate rainwater in either storage basins or planted areas. When assessing the health risks of drinking rainwater, consider the path taken by the raindrop through a watershed into a reservoir, through public drinking water treatment and distribution systems to the end user.

Being the universal solvent, water absorbs contaminants and minerals on its travels to the reservoir. While in residence in the reservoir, the water can come in contact with all kinds of foreign materials: oil, animal wastes, chemical and pharmaceutical wastes, organic compounds, industrial outflows, and trash. It is the job of the water treatment plant to remove harmful contaminants and to kill pathogens.

Unfortunately, when chlorine is used for disinfection, it also degrades into disinfection byproducts, notably trihalomethanes, which may pose health risks. In contrast, the raindrop harvested on site will travel down a roof via a gutter to a storage tank. Before it can be used for drinking, it will be treated by a relatively simple process with equipment that occupies about 9 cubic feet of space.

Rainwater harvesting can reduce the volume of storm water, thereby lessening the impact on erosion and decreasing the load on storm sewers. Decreasing storm water volume also helps keep potential storm water pollutants, such as pesticides, fertilizers, and petroleum products, out of rivers and groundwater. But along with the independence of rainwater harvesting systems comes the inherent responsibility of operation and maintenance. For all systems, this responsibility includes purging the firstflush system, regularly cleaning roof washers and tanks, maintaining pumps, and filtering water. For potable systems, responsibilities include all of the above, and the owner must replace cartridge filters and maintain disinfection equipment on schedule, arrange to have water tested, and monitor tank levels.

Rainwater used for drinking should be tested, at a minimum, for pathogens. Rainwater harvesting, in its essence, is the collection, conveyance, and storage of rainwater. The scope, method, technologies, system complexity, purpose, and end uses vary from rain barrels for garden irrigation in urban areas, to large-scale collection of rainwater for all domestic uses.

Rainwater harvesting is suitable for all building types ranging from residential to commercial and industrial and can be retro-fitted to existing buildings or integrated into new building designs. Collected rooftop water is typically used for non-potable (non-drinking water) demands, but can be treated to drinking water standards and used for potable (drinking water) demands. While many common water uses are non-potable, water supplied from the municipal system and wells is generally potable water. Utilising potable water for non-potable needs wastes resources and can place unneeded strain on local water treatment plants.

Rainwater harvesting is an appropriate water supply and stormwater solution for residential, commercial, industrial, and agricultural applications. All systems consist of the same basic components: a collection surface (only roof surfaces are addressed in this document), a conveyance system, pre-tank treatment, water storage and distribution. Harvested rainwater used for indoor use will also often include additional treatment.

1.5.1. Residential Uses

Residential systems can be designed for non-potable and potable needs. If a potable water source is available, it is recommended that the rainwater harvesting system be used only for non-potable needs like toilet flushing, laundry washing, and landscape irrigation. These non-potable uses account for 78% of total household water use, with outdoor use alone accounting for 59% of household use. Because outdoor use is such a dominant residential water use, bringing harvested rainwater indoors for non-potable use is often not cost-effective, particularly in retro-fit situations.

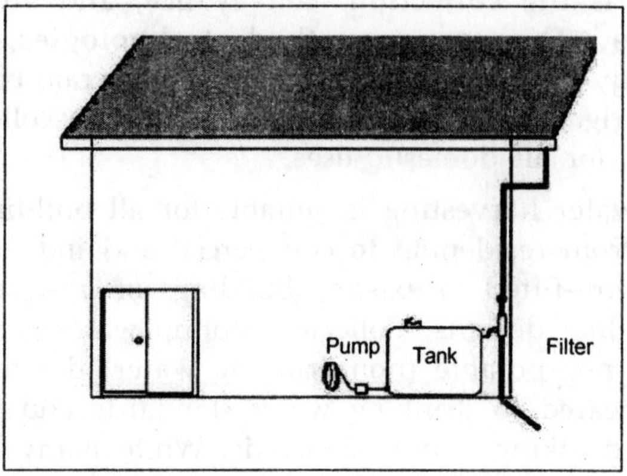

Figure 1.1. A small residential non-potable system, suitable for irrigation purposes.

The cost of a rainwater harvesting system is increased when the water is used indoors because plumbing for potable and non-potable rainwater sources must be separate systems. Municipal or well water can serve as a backup source of water if the rainwater runs dry. An air gap or backflow preventer is necessary between rainwater and municipal water supplies leading to the storage tank to prevent cross contamination.

Rainwater harvesting can serve as an alternative water source for rural homes where municipal water is not available and well drilling has not been successful. Rainwater can also supplement well or municipal water supplies to reduce demand on these supplies to serve non-potable demands in and around the home.

1.5.2. Commercial Uses

Commercial rainwater harvesting systems can supply non-potable water needs. As in residential settings, the majority of commercial use is for non-potable needs. For example, eighty-seven percent of water use in an office building is for restrooms, cooling and landscape use. In hotels, fifty-one percent of water use is for restrooms alone. Because many commercial facilities want a year-round use for stormwater management needs, rainwater use for toilet flushing is common, especially in buildings with high occupancy loads. The soft rainwater is also beneficial for cleaning purposes as less detergent is needed. Water demand, roof size, and available onsite storage should be considered when sizing a collection tank. Figure 1.2 shows a commercial or large residential above ground rainwater harvesting system.

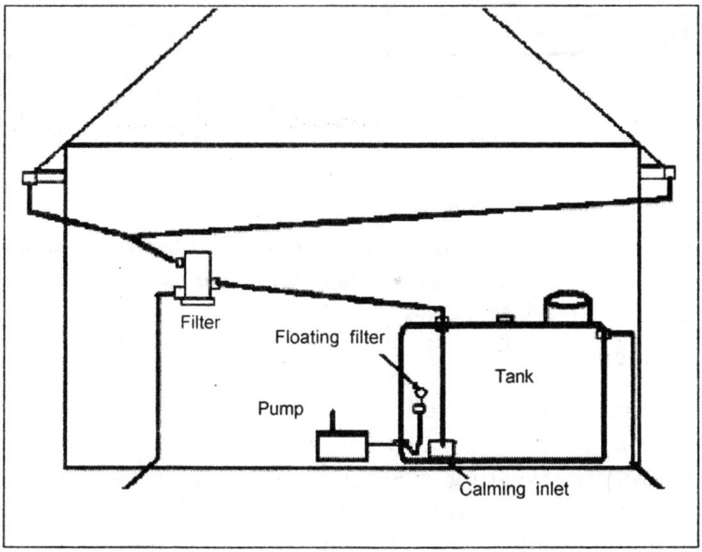

Figure 1.2. Commercial or large residential above ground non-potable rainwater harvesting system

Fig. 1.3 shows a commercial or large residential below ground system. A larger filter is needed to filter water from a larger roof area. In nearly all cases, a submersible or jet pump is necessary to pump water for indoor use. The addition of a calming inlet allows water to enter the tank without disturbing the important sediment layer on the bottom of the tank. A floating filter serves as the elevated uptake point for the pump system. If rainwater is used for potable needs in a commercial setting, the system will typically qualify as a public water system.

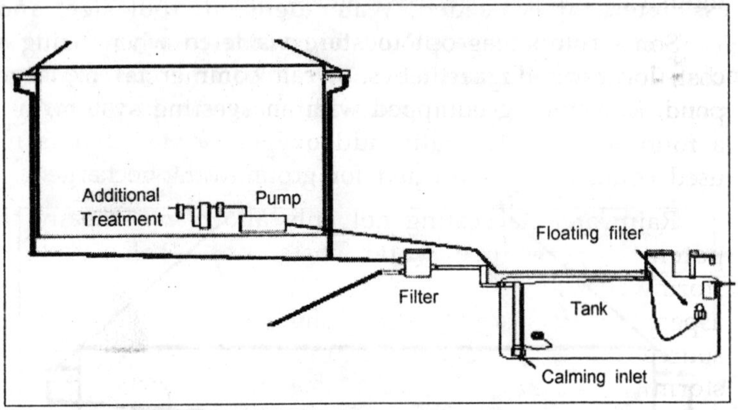

Figure 1.4. Large residential/commercial rainwater harvesting system with belowground tank for indoor use.

1.5.3. Industrial Uses

Industrial buildings with flat roofs can frequently benefit from incorporating a siphonic roof drainage system with a rainwater harvesting system. A siphonic roof drainage system does not require pitched pipes which can reduce or eliminate underground piping. Siphonic roof drainage systems also use smaller diameter pipes. The pipes have to be full to create a siphon while pipes in conventional roof drainage systems are often only half full. This difference allows siphonic roof drainage systems to carry the same volume of water in smaller pipes which reduces the cost of materials. Installation costs can be reduced by 30 to 40% due to the lower excavation, backfill and trenching costs.

Harvested rooftop rainwater can be used indoors to flush toilets, clean floors, and wash linens or as irrigation. As in the commercial setting, the soft rainwater is beneficial for cleaning purposes as less detergent is needed, which not only saves money but also reduces the amount of detergent released into the environment. In addition, cooling and process water often account for 80-90% of industrial water use. Harvested rainwater can be used for cooling in all industries and for process water in many industries.

Some companies opt to store water in a pond due to cost, location and aesthetics. If rainwater is diverted to a pond, it should be equipped with an aerator in the form of a fountain to continually add oxygen. Water that is not used could also be directed for groundwater recharge.

Rainwater harvesting not only allows a company the potential to reduce water costs, but it also reduces stormwater runoff on the site. The stormwater reduction aspect of rainwater harvesting holds this alternative water source above others because rather than contributing to stormwater runoff or energy consumption, it is reducing pollution and protecting local waterways. The ability of this system to reduce stormwater runoff can also be leverage for obtaining building permits.

1.5.4. Agricultural Uses

Rainwater harvesting is ideal for farm animal drinking water and agricultural and landscape irrigation, as it is salt free, easily attained, and reduces groundwater depletion and pumping from local streams. Rainwater can be collected from rooftop surfaces like barns, clubhouses, greenhouses and equipment storage buildings. During summer months, we often experiences quick, hard rainfalls. Such rain events produce too much water in too short a time period for the ground to absorb, which results in most of the rainwater being lost to runoff. Rainwater harvesting systems are capable of collecting rainwater from nearby roof surfaces during such rainfalls. Therefore, the heavy storm's rainwater can be reapplied to the field at a suitable rate to promote water infiltration. Frequently, livestock buildings allow rainwater from the roofs to flow into areas occupied by animals. This situation allows rainwater from these facilities to become contaminated by fecal matter resulting in a greater volume of manure slurry that must be managed. If rainwater from these facilities is harvested, it can be used for watering livestock and washing down facilities and animals.

Harvesting rainwater from these facilities can reduce the amount of waste water created as well as decreasing the demand on wells and other water supplies. Automatic watering troughs are an efficient way to provide clean drinking water to livestock. Most of these units operate using pressure valves and can be fed by an above ground, gravity fed rainwater harvesting system. The roofs of loafing sheds or other agricultural structures can be the source of this water. By decreasing pipe diameter of the pipe feeding the trough, water pressure is increased to a level which can successfully operate the trough's valves. This type of rainwater harvesting system allows for a quality water supply at remote locations.

1.5.5. Fire Suppression

Rainwater harvesting offers alternatives to municipally supplied water for fire suppression. Harvested rainwater can be directed to interior sprinkler systems and used in the advent of a building fire.

Fire suppression can go beyond indoor sprinkler systems and protect buildings from forest fires. Stored water flows backwards into the gutter system and overflows the gutters to form a shield of water. While forest fires are not as common in India as they are in the arid west, rainwater could serve as protection for some homes located in heavily forested areas in the advent of a forest fire. Another alternative is to collect rainwater for fire hydrants. Rooftop and street runoff can be directed to an underground tank connected to a fire hydrant. This prevents the reliance on potable water to fight fires and can reduce connection costs, especially in areas outside the main water distribution grid. Rainwater harvesting systems can also be used to meet fire suppression water supply requirements in remote areas.

1.5.6. Irrigation

Because irrigation is such a large water use across all building types, reducing the quantity of water needed for irrigation is an important step. Approximately 34% of all water use in the United States is for irrigation. Fifty percent of that water can be lost to evaporation and runoff. Therefore, when coupling rainwater harvesting systems with irrigation, home/business owners should take extra steps to ensure the harvested rainwater is applied as efficiently as possible across the landscape. Applying more water than needed can cause runoff and nonpoint source pollution, thus canceling out the conservation and reduction in nonpoint source pollution benefits of rainwater harvesting.

1.6. Constraints in Rainwater Harvesting

Rainwater harvesting is practical only when the volume and frequency of rainfall and size of the catchment surface can generate sufficient water for the intended purpose. From a financial perspective, the installation and maintenance costs of a rainwater harvesting system for potable water cannot compete with water supplied by a central utility, but is often cost-competitive with installation of a well in rural settings. With a very large catchment surface, such as that of big commercial building, the volume of rainwater, when captured and stored, can cost-effectively serve several end uses, such as landscape irrigation and toilet flushing.

Some commercial and industrial buildings augment rainwater with condensate from air conditioning systems. During hot, humid months, warm, moisture-laden air passing over the cooling coils of a residential air conditioner can produce 10 or more gallons per day of water. Industrial facilities produce thousands of gallons per day of condensate. An advantage of condensate capture is that its maximum production occurs during the hottest month of the year, when irrigation need is greatest. Most systems pipe condensate into the rainwater cistern for storage.

The depletion of groundwater sources, the poor quality of some groundwater, high tap fees for isolated properties, the flexibility of rainwater harvesting systems, and modern methods of treatment provide excellent reasons to harvest rainwater for domestic use.

References

Frasier, Gary, and Lloyd Myers. *Handbook of Water Harvesting*. Washington D.C.: U.S. Dept. of Agriculture, Agricultural Research Service, 1983.

Geerts, S., Raes, D. "Deficit irrigation as an on-farm strategy to maximize crop water productivity in dry areas". *Agric. Water Manage* 96, 1275-1284. 2009.

Gould, John, and Erik Nissen-Peterson. *Rainwater Catchment Systems*. UK: Intermediate Technology Publications, 1999.

Hemenway, Toby. *Gaia's Garden: A Guide to Home-Scale Permaculture*. Vermont: Chelsea Green Publishing Company, 2000.

Lowes, P. "The Water Decade: Half Time". in in John Pickford (ed.). *Developing World Water*. London: Grosvenor Press International. pp. pp 16-17. ISBN 0-946027-29-3. 1987.

Ludwig, Art. *Create an Oasis With Greywater: Choosing, Building, and Using Greywater Systems*. California: Oasis Design, 1994.

Pacey, Arnold, and Adrian Cullis. *Rainwater Harvesting*. UK: Intermediate Technology Publications, 1986.

2 Basic Components of Rainwater Harvesting System

Today, scarcity of good quality water has become a major cause of concern. However rainwater which is pure and of good quality is lost as run off. The collection and storage of rainwater for later productive use is defined as rainwater harvesting. It is simply collecting, storing and purifying the naturally soft and pure rainfall that falls upon your roof. In a residential or small-scale application, rainwater harvesting can be as simple as channelling rain running off an unguttered roof to a planted landscape area via contoured landscape. To prevent erosion on sloped surfaces, a termed concave holding area down slope can store water for direct use by turfgrass or plants. More complex systems include gutters, pipes, storage tanks or cisterns, filtering, pump(s), and water treatment for potable use.

Any rainwater harvesting system has three components: Catchment, Conveyance, and Storage. There are two classes of rainwater harvesting systems:

1. Systems which collect roof runoff for household use.
2. Systems which use in field or adjoining catchment to provide supplemental irrigation for agriculture.

The six basic components of a rainwater harvesting system include:

1. *Catchment surface*: the collection surface from which rainfall runs off.
2. *Gutters and downspouts*: channel water from the roof to the tank.

2. *Leaf screens, first-flush diverters, and roof washers*: components which remove debris and dust from the captured rainwater before it goes to the tank.
3. *Storage*: cisterns or tanks where collected rainwater is securely stored – i.e. insect proof.
4. *Delivery system*: gravity-fed or pumped to the end use.
5. *Treatment/purification*: for potable systems, filters and other methods to make the water safe to drink.

The basic components of the domestic rainwater harvesting system is shown in Figure 2. 1.

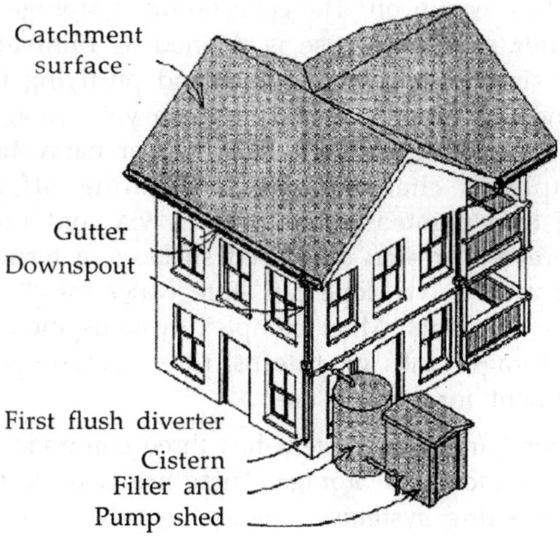

Figure 2.1. Typical rainwater harvesting installation

2.1. Catchment Surface

The roof of a building or house is the obvious first choice for catchment. For additional capacity, an open-sided barn – called a rain barn or pole barn – can be built. Water tanks and other rainwater system equipment, such as pumps and filters, as well as vehicles, bicycles, and gardening tools, can be stored under the barn. Water quality from different roof catchments is a function of the type of roof material, climatic conditions, and the surrounding environment.

When rain falls on a roof, it can pick up contaminants both from the roofing material and material deposited on the roof surface. Material on the roof surface can be controlled some by cleaning the roof and trimming trees back near the roof, but many of the sources of material deposited on the roof, such as nearby industrial facilities, are outside of the rainwater harvesting system designer's hands. Roofing material is, however, a design choice and can have a significant effect on the quality of harvested rainwater. Chemical reactions on the roof are often rapid because of the acidity of rainfall and the high temperatures on many rooftops. These reactions make the choice of roofing material an important consideration in designing a rainwater harvesting system, particularly for potable uses. The choice of roofing material will also affect the quantity of rainwater available as some materials absorb water.

2.1.1. Metal Roofs

Metal roofs, with the exception of copper and roofs with lead components are recommended for rainwater harvesting. Metal roofs have many advantages for rainwater harvesting, including a smooth surface and a high runoff coefficient. Unfortunately, metals are also leached from many of these roofs. Aluminum and zinc are often found in runoff from roofs containing these metals. While presence of these metals in a water supply may have aesthetic effects (effects on color from aluminum and effects on odor and taste from zinc), they have not been identified as health risks. Ingestion of copper, however, can cause gastrointestinal symptoms and possibly kidney and liver damage with long-term exposure. It is identified that copper is a primary drinking water contaminant with a Maximum Contaminant Level Goal of 1.3 ppm. Copper concentration in runoff from copper roofs can be as high as 12 ppm, even when background rainwater concentrations are low. Lead flashing, sometimes used on metal roofs, also presents a significant health risk. Even though lead flashing generally represents only a small percentage of the surface area of a roof, it can still contribute a significant quantity of lead to the run-off causing lead concentrations of up to 0.42 mg/l Because of the potential for contamination, copper roofs and roofs with lead components should not be used for potable uses, fruit/vegetable gardening, or pool filling. Other metal roofs, however, are excellent choices for rainwater harvesting systems.

2.1.2. Asphalt Shingle

Asphaltic roofing material cannot be used as a catchment surface for cisterns if the water will eventually be sent to an on-site septic field.Asphalt shingle roofing has been identified as a source of both lead and mercury. However, the specific composition of asphalt shingles varies widely. For retro-fit situations with an asphalt roof, preliminary water quality sampling of runoff from the roof is likely the best way to determine appropriate end uses of the water. For new construction, roofing materials other than asphalt shingle are generally preferable.

2.1.3. Wood Shingle

Because of a low runoff coefficient and many of the products used to treat wood, wood shingle roofs are not recommended for rainwater harvesting. Chromated copper arsenate (CCA) treated wood and untreated wood have both been shown to contribute arsenic and copper in measurable quantities to runoff. Wood shingles can also reduce the pH of runoff, making it more acidic.

2.1.4. Cement tile/terra Cotta Tile

Tile roofs should be used with caution in rainwater harvesting systems. Semi-porous roof surfaces like terra cotta tiles absorb some of the water, thus reducing the system's water collection efficiency. In addition, tiles have been shown to contribute significant quantities of lead, copper and cadmium to runoff. Finally, the uneven surfaces and absorbed water in the tiles may create an ideal growing environment for bacteria and algae. Asbestos

Asbestos/cement shingles and tiles were common roofing materials prior to widespread concern and regulation because of the health effects of asbestos. While the primary health risk of asbestos is inhalation of fibers which can lead to respiratory problems and cancer, some studies have found statistically significant increases in cancer rates, particularly gastrointestinal cancers, in areas with asbestos in the drinking water supply. Little research has been done on the asbestos concentration in runoff from roofing material made with asbestos, but the available research does indicate that concentrations may exceed drinking water standards. Out of precaution, rainwater harvested from roofs made of asbestos materials is not recommended for any use with significant human contact.

Green roof Vegetated roof surfaces with a soil base absorb most of the water that falls on the roof and only 10% to 20% of the runoff is collected. The collected runoff is typically a brown color, thus only suitable for landscape irrigation. However, up to 30% of rainfall runoff from vegetated roofs with a gravel base is collected. This water is also a clear color so it is suitable for both indoor and outdoor use.

Membrane roof Membrane roofs create an ideal surface for rainwater harvesting. These roofs have a high runoff coefficient and have not been shown to add harmful contaminants to the harvested rainwater.

2.1.5. Roof Slope

Roofs with a pitch work best for rainwater harvesting, as water is easily moved through gravitational force. Since organic matter can build up between rain events, a steeper roof also allows water to move more efficiently and quickly across the surface, which helps clean the roof surface.

2.2. Gutters and Downspouts

A rainwater harvesting system is compatible with any gutter and downspout system, but research has shown that some design choices can improve the functioning of the system. Gutters have two main functions: capturing water from the roof (interception) and conveying this water to the downspouts (conveyance). Both of these functions are affected by gutter shape and gutter slope. Research shows that guttering systems should be pitched 0.5% for 2/3 of the length and 1% for the remaining 1/3 of the length and ideally a semi-circular or trapezoidal shape (Fig. 2.2) for optimal interception and conveyance.

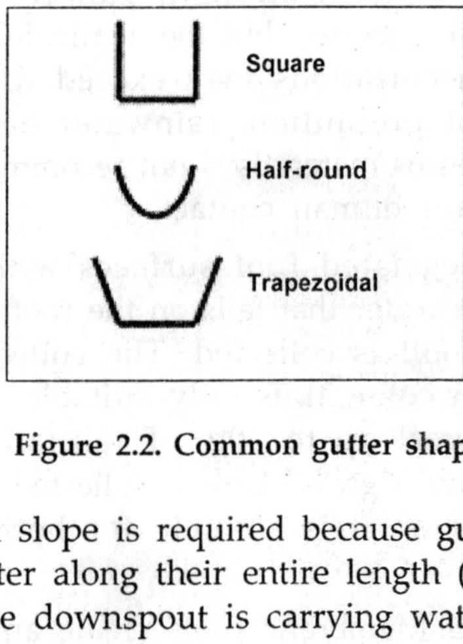

Figure 2.2. Common gutter shapes

The change in slope is required because gutters continue to collect rainwater along their entire length (i.e. the part of a gutter near the downspout is carrying water from a bigger roof area and therefore more water than farther from the downspout). Increasing the slope lets the same size gutter carry a greater volume of water.

Half-round and trapezoidal gutters are suggested because they are able to drain a greater roof area (i.e. carry more water) with the same amount of material used to make the gutter (the same gutter perimeter). Assuming that the price of gutters is dependent on the amount of material used to make them, the half-round and trapezoidal gutters should be more cost-effective and water velocity should be higher leading to more efficient draining and removal of debris.

Gutter systems should remain free from debris at all times. This ensures water moves freely from roof surfaces to the storage tank. Installing covered gutters or adding guards to existing gutters is ideal to prevent debris buildup and clogging.

2.2.1. Roof Drains

While rainwater harvesting systems can be integrated with any roof drain system, siphonic roof drains (Fig. 2.3) present a number of advantages. Siphonic roof drain systems generally require fewer, and smaller, downpipes and less underground piping.

Figure 2.3. Siphonic roof drain.

The system designer can frequently locate all downpipes on one side of the building. When siphonic roof drains are used in conjunction with rainwater harvesting, this means that the downpipes can be located near the cistern. When conventional roof drains are used instead of siphonic, pipes often have to be brought longer distances (often either around or under the building). This not only requires additional pipe and site work, but because of the required slope on the pipe, can result in deeper, and therefore more expensive, burial of the cistern (Fig. 2.4).

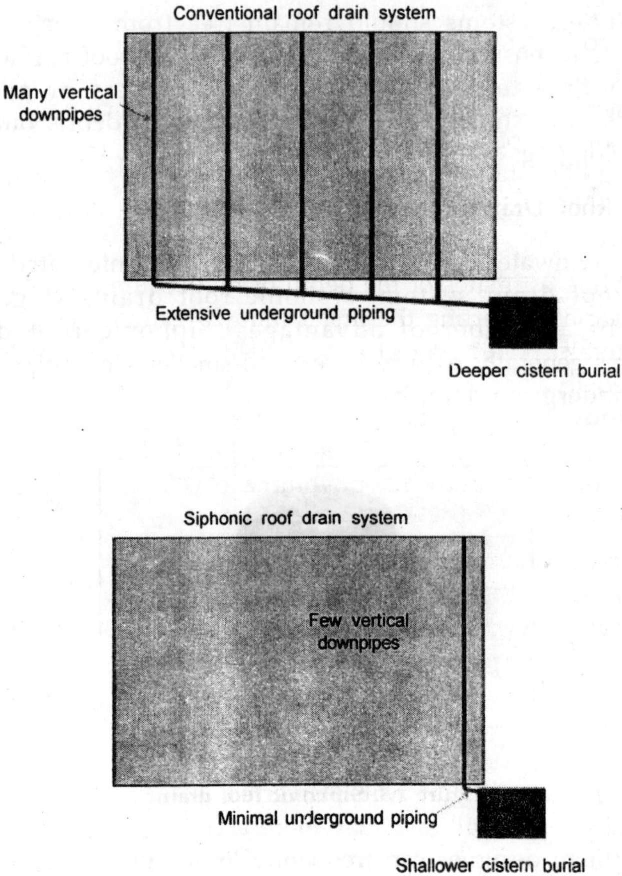

Figure 2.4. Schematic representations of a conventional roof drain system and a siphonic roof drain system.

2.2.2. First Flush Diversion

The initial runoff from a roof surface, called the first flush generally rinses the roof, leading to cleaner water as the rainfall continues. For example, both iron and zinc concentrations in runoff from a polyester roof were four times higher during the first 0.08? of rainfall than the steady state concentration later during the same rain event. This trend of cleaner water later in a rain event is true for sediment , metals , bacteria , and even pesticides. Because most of the possible contaminants are rinsed from the roof during the start of a rain event, diverting the first flush from the tank should greatly improve the stored water quality.

While estimates of the depth of rain required to rinse the roof varies, diverting the first 1 mm (25 gallons per 1,000 ft^2) of rainfall should protect water quality. The depth of rainfall needed to rinse the roof and bring contaminant levels down to steady state depends on the intensity of the rainfall and the type and condition of the roof. First flush diversion can be accomplished by fixed volume diverters or first flush filters.

Fixed volume typically filled a holding chamber with rainwater at the start of an event and once the chamber is full, rainwater is then diverted into the tank. The chamber periodically must be cleaned to rid the unit of clogging debris, which decreases the unit's efficiency and impedes the collected rainwater quality. First flush filters, however, divert rainwater until the stainless steel mesh is wet. In this way, the quantity of water diverted is proportional to the intensity of rainfall and may more accurately match the true first flush volume.

2.2.3. Pre-tank Filters

Pre-tank filters protect the quality of rainwater stored in the tank by preventing the introduction of debris. A study of rainwater harvesting tanks in Ontario showed generally improved cistern water quality in a system with a vortex pre-tank filter compared to systems with only first flush diverters or no pre-tank treatment. The logic behind pre-tank filtration is simple. A build-up of organic debris in the tank would lead to greater decomposition resulting in low oxygen levels and a build up nutrients in the tank. Low oxygen conditions can lead to the development of odors as well as favor the growth of harmful bacteria in the tank.

Filters collect rainwater through a straining action, which occurs through some form of screen or mesh between the rainwater from the roof and the clean rainwater outlet to the tank. Depending on the design of the filter, rainwater quality can be impeded if leaves are allowed collect on the screen or mesh. Not only is the filter's efficiency is reduced, decaying leaves leach contaminants into the stored rainwater.

Filters equipped with made screens or mesh that is self-drying and prevent microbial growth are the best at protecting rainwater quality and filter efficiency. Filter fabric should dry between rainfall events to prevent algae and biofilm growth, which could block the fabric pores. Also, fabrics should be made of stable materials that do not change shape and can withstand temperature changes, ice formation, and frost. Stainless steel is considered the best filter fabric because it can withstand all weather conditions, even ice formation and frost, is self cleaning and self drying, maintains shape, and does not rust, thus reducing contamination likelihood.

There are few filters on the market that utilise both the first flush and straining mechanisms to harvest rainwater. The goal of a high-quality filter is to not only eliminate contaminants but also supply oxygen to water during the filtration process. An advanced filter does not restrict the diameter of the gutter and is positioned either vertically connected to the gutter system or horizontally connected to the downspouts.

Modern filters require low maintenance and cleaning and can efficiently collect more than 90% of filtrated rainwater. To ensure the effectiveness of the filter, the appropriate filter should be paired with the appropriate roof area. Also, utilising high quality filters ensures water is sufficiently filtered, oxygenated and directed to storage tanks. High quality filters need inspection only a few times a year and should last the lifespan of the building. Purchasing sustainable products for rainwater harvesting further emphasises the environmental impact by conserving water, energy, and production resources. Research should be conducted to ensure high quality products are integrated into a building's rainwater harvesting design.

2.3. Leaf Screens

To remove debris that gathers on the catchment surface, and ensure high quality water for either potable use or to work well without clogging irrigation emitters, a series of filters are necessary. Essentially, mesh screens remove debris both before and after the storage tank. The defense in keeping debris out of a rainwater harvesting system is some type of leaf screen along the gutter or in the downspout.

Depending upon the amount and type of tree litter and dust accumulation, the homeowner may have to experiment to find the method that works best.

Leaf screens must be regularly cleaned to be effective. If not maintained, leaf screens can become clogged and prevent rainwater from flowing into a tank. Built-up debris can also harbor bacteria and the products of leaf decay. Leaf guards are usually ¼-inch mesh screens in wire frames that fit along the length of the gutter. Leaf guards/screens are usually necessary only in locations with tree overhang. Guards with profiles conducive to allowing leaf litter to slide off are also available.

The funnel-type downspout filter is made of PVC or galvanised steel fitted with a stainless steel or brass screen. This type of filter offers the advantage of easy accessibility for cleaning. The funnel is cut into the downspout pipe at the same height or slightly higher than the highest water level in the storage tank.

Strainer baskets are spherical cage-like strainers that slip into the drop outlet of the downspout.

A cylinder of rolled screen inserted into the drop outlet serves as another method of filtering debris. The homeowner may need to experiment with various grid sizes, from insect screen to hardware cloth.

Filter socks of nylon mesh can be installed on the PVC pipe at the tank inflow.

2.4. First-Flush Diverters

A roof can be a natural collection surface for dust, leaves, blooms, twigs, insect bodies, animal feces, pesticides, and other airborne residues. The firstflush diverter routes the first flow of water from the catchment surface away from the storage tank. The flushed water can be routed to a planted area. While leaf screens remove the larger debris, such as leaves, twigs, and blooms that fall on the roof, the first-flush diverter gives the system a chance to rid itself of the smaller contaminants, such as dust, pollen, and bird and rodent feces.

The simplest first-flush diverter is a PVC standpipe. The standpipe fills with water first during a rainfall event; the balance of water is routed to the tank. The standpipe is drained continuously via a pinhole or by leaving the screw closure slightly loose. In any case, cleaning of the standpipe is accomplished by removing the PVC cover with a wrench and removing collected debris after each rainfall event.

There are several other types of firstflush diverters. The ball valve type consists of a floating ball that seals off the top of the diverter pipe when the pipe files with water.

Opinions vary on the volume of rainwater to divert. The number of dry days, amount of debris, and roof surface are all variables to consider.

One rule of thumb for first-flush diversion is to divert a minimum of 10 gallons for every 1,000 square feet of collection surface. However, first-flush volumes vary with the amount of dust on the roof surface, which is a function of the number of dry days, the amount and type of debris, tree overhang, and season.

A preliminary study by Rainwater Harvesting and Waste Water Systems Pty Ltd., a rainwater harvesting component vendor in Australia, recommends that between 13 and 49 gallons be diverted per 1,000 square feet.

The primary reason for the wide variation in estimates is that there is no exact calculation to determine how much initial water needs to be diverted because there are many variables that would determine the effectiveness of washing the contaminants off the collection surface, just as there are many variables determining the make up of the contaminants themselves. For example, the slope and smoothness of the collection surface, the intensity of the rain event, the length of time between events (which adds to the amount of accumulated contaminants), and the nature of the contaminants themselves add to the difficulty of determining just how much rain should be diverted during first flush. In order to effectively wash a collection surface, a rain intensity of one-tenth of an inch of rain per hour is needed to wash a sloped roof. A flat or near-flat collection surface requires 0.18 inches of rain per hour for an effective washing of the surface.

The recommended diversion of first flush ranges from one to two gallons of first-flush diversion for each 100 square feet of collection area. If using a roof for a collection area that drains into gutters, calculate the amount of rainfall area that will be drained into every gutter feeding your system. Remember to calculate the horizontal equivalent of the "roof footprint" when calculating your catchment area. If a gutter receives the quantity of runoff that require multiple downspouts, first-flush diversion devices will be required for each downspout.

2.5. Roof Washers

The roof washer, placed just ahead of the storage tank, filters small debris for potable systems and also for systems using drip irrigation. Roof washers consist of a tank, usually between 30-and 50-gallon capacity, with leaf strainers and a filter. One commercially available roof washer has a 30-micron filter. (A micron, also called a micrometer, is one-millionth of a meter. A 30-micron filter has pores about one-third the diameter of a human hair.)

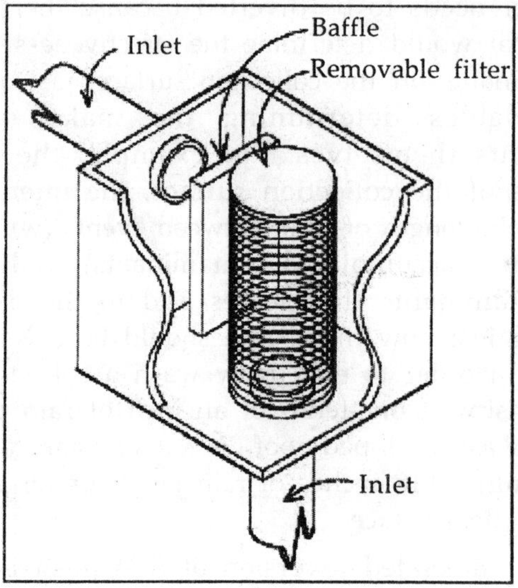

Figure 2.5. Box Roof Washer

All roof washers must be cleaned. Without proper maintenance they not only become clogged and restrict the flow of rainwater, but may themselves become breeding grounds for pathogens. The box roof washer is a commercially available component consisting of a fiberglass box with one or two 30-micron canister filters (handling rainwater from 1,500- and 3,500-square-foot catchments, respectively).

The box is placed atop a ladder-like stand beside the tank, from which the system owner accesses the box for cleaning via the ladder. In locations with limited drop, a filter with the canisters oriented horizontally is indicated, with the inlet and outlet of the filter being nearly parallel.

2.6. Storage Tanks

The storage tank is the most expensive component of the rainwater harvesting system.The size of storage tank or cistern is dictated by several variables: the rainwater supply (local precipitation), the demand, the projected length of dry spells without rain, the catchment surface area, aesthetics, personal preference, and budget.

A myriad of variations on storage tanks and cisterns have been used over the centuries and in different geographical regions: earthenware cisterns in prebiblical times, large pottery containers in Africa, above-ground vinyl-lined swimming pools in Hawaii, concrete or brick cisterns in the central United States, galvanised steel tanks and attractive site-built stone-and-mortar cisterns.

For purposes of practicality, this chapter will focus on the most common, easily installed, and readily available storage options, some still functional after a century of use.

Storage tanks must:

- be opaque, either upon purchase or painted later, to inhibit algae growth.
- never have been used to store toxic materials.
- be covered and vents screened to discourage mosquito breeding.
- be accessible for cleaning.

Tank selection is based on three main criteria: size, location and material. Tank size is always dependent on the roof area and the anticipated use of the water. However, size decisions may also be based on availability of space on the site, and the availability of a back-up water supply. Tank size selection will affect possible tank locations and location and size will then help in the selection of tank materials.

2.6.1. Tank Sizing

While rainwater harvesting has a number of environmental benefits, it must also be economically viable to enter the mainstream of building practices. Because the tank is often the most expensive component of a rainwater harvesting system, decision-making on tank size can have a strong impact on the payback period and economic feasibility of rainwater harvesting. The first step in selecting a tank size is choosing a tank selection criterion:

— If the back-up water supply is of poor quality or low quantity, does the rainwater harvesting system need to meet the demand as close to 100% of the time as possible?

— Does the tank need to be the most cost-effective size possible?

— Does the tank need to meet stormwater demands?

In all cases, the tank should overflow occasionally so that floating debris can be skimmed from the water surface. Once the tank sizing criterion has been selected, the most accurate way to select the appropriate tank size is to model the system. The volume of rainwater collected from a surface can be calculated as:

$$\text{Volume of rainwater(gal)} = \text{Area of roof(ft}^2\text{)} \times \text{Collection efficiency} \times \text{Depth of rainfall(in)} \times 0.62$$

The collection efficiency is based on both the roof type and the efficiency of the filters and 0.62 is a conversion factor from f^{t2}-in to gallons. With this equation and estimates of how much water will be used, the system can be modeled to determine the best tank size. Larger tanks are able to supply a greater percentage of demand because they can capture more rain from large events. However, there is often a point of diminishing returns beyond which increasing the tank size provides only a marginal benefit (Fig. 2.6).

Using monthly averages of rainfall can lead to significant errors in tank sizing. Imagine two cities; each receives 4 inches of rain per month. If City A has only one storm a month and all 4 inches of rain fall in one day, a large storage tank will be needed. If City B receives a little rain each day, only a small storage tank will be needed. For example, for a 10,000 ft2 footprint office building, in City A, the rainwater harvesting storage tank would need to hold over 20,000 gallons, while in City B it could be less than 900 gallons to capture all of the rainfall if water was used every day. Researchers have modeled rainwater systems with time intervals as small as 15 minutes but have found that daily modeling is generally sufficient.

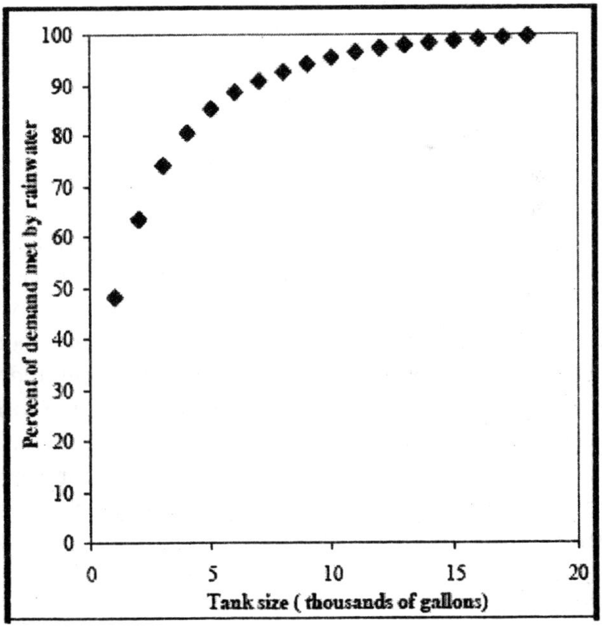

Figure 2.6. Sample graph showing how storage tank size affects the amount of water supplied by a rainwater harvesting system. As tank size increases, so does the percent of demand met, though the increase is very small at large tank sizes.

2.6.2. Tank Location

Tanks can be placed either above or below ground. For both options, water should be able to gravity-feed to the tank except in rare circumstances. Locating the tank near the building and the water use reduces the amount of pipe and site work necessary as well as pump demands. Though no hard and fast rules govern the decision of aboveground or belowground tanks, in general, once the storage volume exceeds 10,000 gallons or multiple roof drains or downspouts are being used, belowground storage is often the most viable option. Beyond this, tank location is dependent on aesthetics, climate and soil conditions. Some prefer not to see the storage tank and opt to bury it. However, placing tanks underground adds to the installation costs and may be limited in areas where soil is especially rocky or areas with a high water table. When tanks are installed below ground, water is maintained at a cool temperature and light is blocked, which reduces the chances of bacterial growth.

2.6.3. Tank Materials

In general, any water storage tank can be used in a rainwater harvesting system, though a few features of common tank materials must be taken into account. Similar to metal roofs, unlined metal tanks may leach some metals into the stored water which may make water unsuitable for drinking. Concrete tanks provide an advantage over other tanks systems because they neutralise the acidity of harvested rainwater. Some tanks are not suitable for certain climatic areas; wooden tanks are not recommended for hot, dry locations.

2.6.3.1. Fiberglass

Fiberglass tanks are built in standard capacities from 50 gallons to 15,000 gallons and in both vertical cylinder and low-horizontal cylinder configurations. Fiberglass tanks under 1,000 gallons are expensive for their capacity, so polypropylene might be preferred. Tanks for potable use should have a USDAapproved food-grade resin lining and the tank should be opaque to inhibit algae growth.

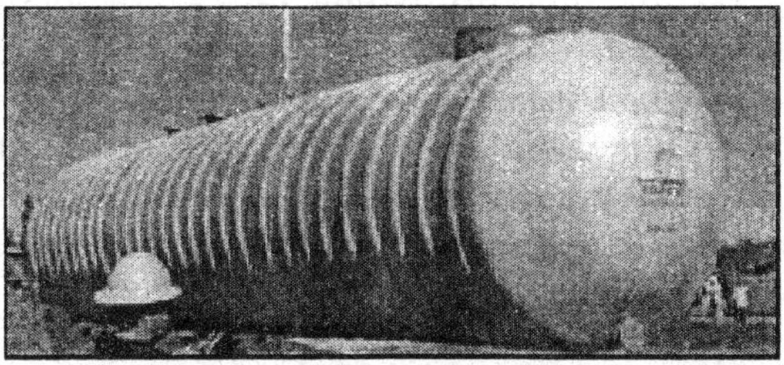

Figure 2.7. Fiberglass tank

The durability of fiberglass tanks has been tested and proven, weathering the elements for years in oil fields. They are easily repaired. The fittings on fiberglass tanks are an integral part of the tank, eliminating the potential problem of leaking from an aftermarket fitting.

2.6.3.2. *Polypropylene*

Polypropylene tanks are commonly sold at farm and ranch supply retailers for all manner of storage uses. Standard tanks must be installed above ground.

Figure 2.8. Polypropylene tank

For buried installation, specially reinforced tanks are necessary to withstand soil expansion and contraction. They are relatively inexpensive and durable, lightweight, and long lasting. Polypropylene tanks are available in capacities from 50 gallons to 10,000 gallons. Polypropylene tanks do not retain paint well, so it is necessary to find off-theshelf tanks manufactured with opaque plastic. The fittings of these tanks are aftermarket modifications. Although easy to plumb, the bulkhead fittings might be subject to leakage.

2.6.3.3. Wood

For aesthetic appeal, a wood tank is often a highly desirable choice for urban and suburban rainwater harvesters. Wood tanks, similar to wood water towers at railroad depots, were historically made of redwood.

Figure 2.9. Wooden tank

Modern wood tanks are usually of pine, cedar, or cypress wrapped with steel tension cables, and lined with plastic. For potable use, a food-grade liner must be used. These tanks are available in capacities from 700 to 37,000 gallons, and are sitebuilt by skilled technicians. They can be dismantled and reassembled at a different location.

2.6.3.4. *Metal*

Galvanised sheet metal tanks are also an attractive option for the urban or suburban garden. They are available in sizes from 150 to 2,500 gallons, and are lightweight and easy to relocate. Tanks can be lined for potable use.

Figure 2.10. Galvanised sheet metal tank

Most tanks are corrugated galvanised steel dipped in hot zinc for corrosion resistance. They are lined with a food-grade liner, usually polyethylene or PVC, or coated on the inside with epoxy paint. The paint, which also extends the life of the metal, must be FDA- and NSF-approved for potability.

2.6.3.5. *Concrete*

Concrete tanks are either poured in place or prefabricated. They can be constructed above ground or below ground. Poured-in-place tanks can be integrated into new construction under a patio, or a basement, and their placement is considered permanent. Concrete may be prone to cracking and leaking, especially in underground tanks in clay soil. Leaks can be easily repaired although the tank may need to be drained to make the repair. Involving the expertise of a structural engineer to determine the size and spacing of reinforcing steel to match the structural loads of a poured-in-place concrete cistern is highly recommended. A product that repairs leaks in concrete tanks, Xypex™, is now also available and approved for potable use.

Figure 2.11. Concrete tank

One possible advantage of concrete tanks is a desirable taste imparted to the water by calcium in the concrete being dissolved by the slightly acidic rainwater. For potable systems, it is essential that the interior of the tank be plastered with a high-quality material approved for potable use.

2.6.3.6. Ferrocement

Ferrocement is a low-cost steel and mortar composite material. For purposes of this chapter, GuniteTM and ShotcreteTM type will be classified as ferrocements. Both involve application of the concrete and mortar under pressure from a gun. Gunite, the dry-gun spray method in which the dry mortar is mixed with water at the nozzle, is familiar for its use in swimming pool construction. Shotcrete uses a similar application, but the mixture is a prepared slurry. Both methods are cost-effective for larger storage tanks.

Figure 2.12. Ferrocement Tank

Tanks made of Gunite and Shotcrete consist of an armature made from a grid of steel reinforcing rods tied together with wire around which is placed a wire form with closely spaced layers of mesh, such as expanded metal lath. A concrete-sand-water mixture is applied over the form and allowed to cure. It is important to ensure that the ferrocement mix does not contain any toxic constituents. Some sources recommend painting above-ground tanks white to reflect the sun's rays, reduce evaporation, and keep the water cool. Ferrocement structures have commonly been used for water storage construction in developing countries due to low cost and availability of materials.

Small cracks and leaks can easily be repaired with a mixture of cement and water, which is applied where wet spots appear on the tank's exterior. Because walls can be as thin as 1 inch, a ferrocement tank uses less material than concrete tanks, and thus can be less expensive. As with pouredin-place concrete construction, assistance from a structural engineer is encouraged.

2.6.3.7. *In-ground Polypropylene*

In-ground tanks are more costly to install for two reasons: the cost of excavation and the cost of a more heavily reinforced tank needed if the tank is to be buried more than 2-feet deep in well-drained soils. Burying a tank in clay is not recommended because of the expansion/contraction cycles of clay soil. For deeper installation, the walls of poly tanks must be manufactured thicker and sometimes an interior bracing structure must be added. Tanks are buried for aesthetic or space-saving reasons.

Table 1. Round Cistern Capacity (Gallons)

Height (feet)	*6-foot Diameter*	*12-foot Diameter*	*18-foot Diameter*
6	1,269	5,076	11,421
8	1,692	6,768	15,227
10	2,115	8,460	19,034
12	2,538	10,152	22,841
14	2,961	11,844	26,648
16	3,384	13,535	30,455
18	3,807	15,227	34,262
20	4,230	16,919	38,069

Table 2.1 provides some values to assist in planning an appropriate-sized pad and cistern to meet your water needs and your available space. Many owners of rainwater harvesting systems use multiple smaller tanks in sequence to meet their storage capacity needs. This has the advantage of allowing the owner to empty a tank in order to perform maintenance on one tank at a time without losing all water in storage. A summary of cistern materials, their features, and some words of caution are provided in Table 2.2 to assist the prospective harvester in choosing the appropriate cistern type. Prior to making your final selection, consulting with an architect, engineer, or professional rainwater installer is recommended to ensure the right choice for your situation.

Table 2.2. Cistern Types

Material	*Features*	*Caution*
Plastics		
Trash cans (20-50 gallon)	commercially available; inexpensive	use only new cans
Fiberglass	commercially available; alterable and moveable	must be sited on smooth, solid, level footing
Polyethylene/polypropylene	commercially available; alterable and moveable	UV-degradable, must be painted or tinted
Metals		
Steel drums (55-gallon)	commercially available; alterable and moveable	verify prior to use for toxics; prone to corrosion an rust;
Galvanized steel tanks	commercially available; alterable and moveable	possibly corrosion and rust; must be lined for potable use
Concrete and Masonry		
Ferrocement	durable and immoveable	potential to crack and fail
Stone, concrete block	durable and immoveable	difficult to maintain
Monolithic/Poured-in-place	durable and immoveable	potential to crack
Wood		
Redwood, fir, cypress	attractive, durable, can be disassembled and moved	expensive

2.6.4. Rain Barrel

One of the simplest rainwater installations, and a practical choice for urban dwellers, is the 50- to 75-gallon drum used as a rain barrel for irrigation of plant beds. Some commercially available rain barrels are manufactured with overflow ports linking the primary barrel to a second barrel. A screen trap at the water entry point discourages mosquito breeding. A food-grade plastic barrel used for bulk liquid storage in restaurants and grocery stores can be fitted with a bulkhead fitting and spigot for garden watering. Other options include a submersible pump or jet pump.

2.6.5. Rainwater Inlet to the Tank

In a healthy rainwater storage tank, sediments settle to the bottom and often incorporated into the biofilm. Many contaminants in roof runoff are associated with particles and higher contaminant concentrations are often found in this sediment layer. Calming inlets, located at the bottom of storage tanks, direct the entering water upwards to prevent disturbing the fine particulate matter and biofilm on the bottom of the tank (Fig. 2.13). The calming inlet also introduces oxygen into the bottom of the tank. Inlets should be spaced to provide oxygenation throughout the tank.

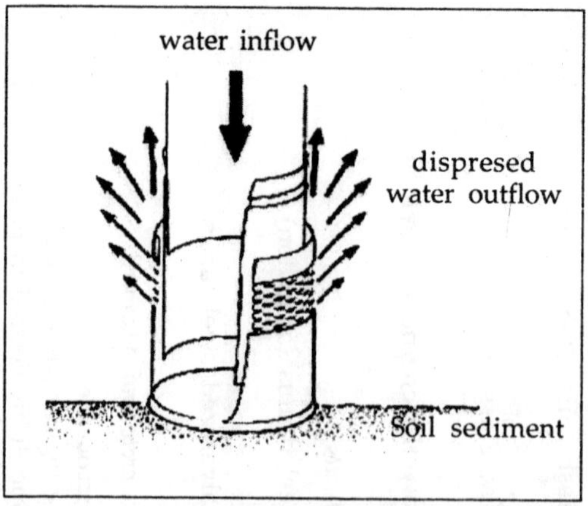

Figure 2.13. Calming inlet

2.6.5.1. Biofilm

Biofilms, layers of bacteria attached to a solid surface with a protective binding often made of a chain of carbohydrates, occur throughout nature, on riverbeds and even on human teeth (plaque is a type of biofilm). While few people like to think about a bacteria layer in their water storage tank, biofilms are an important part of a healthy rainwater harvesting system. Research has shown that biofilms in rainwater systems typically develop from bacteria that are common in soil and the environment, accumulate metals and may remove bacteria from the stored rainwater. In another study, rainwater was pumped through a reactor that contained bacteria. This reactor led to a decrease in concentrations of iron, sulfate and nutrients. Looking at the function of biofilms in water treatment and distribution can help in understanding biofilms in rainwater harvesting systems.

Biofilms (Fig. 2.14) have been used in water and wastewater treatment systems to remove organic matter. However, biofilms in plumbing distribution systems, the pipes that carry water to homes and businesses, increase corrosion in the pipes and can even shelter disease-causing bacteria. These biofilms can be particularly troublesome because they are resistant to most traditional treatment techniques (i.e. chlorination). Using a biofilm in the treatment system, instead of just traditional treatment practices, can decrease the chances of the development of biofilm in the distribution system.

The bacteria in biofilm use organic matter in the water as a food source. If the biofilm in the rainwater tank digests the available organic matter, the biofilm in the distribution system and any bacteria that may live in the water have a reduced food source which will minimise growth. Even with a stable and healthy biofilm, further treatment of the harvested rainwater is recommended if the water will have high contact with humans. This water should also be tested for harmful bacteria.

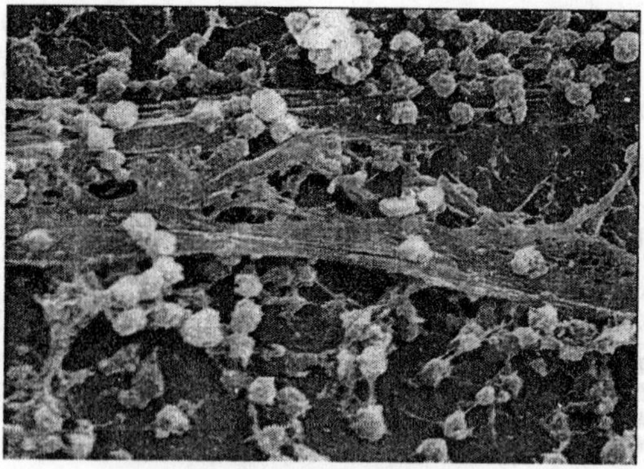

Figure 2.14. Electron micrograph of a biofilm. The round structures are bacteria, the stringy material between them is a polysaccharide that binds them together, making this a biofilm.

The biofilm in a rainwater harvesting storage tank should be protected to ensure high water quality. The tank should never be emptied or cleaned as long as the rainwater passes through a pre-filtering first-flush filter system. Cleaning the tank will kill the biofilm layer.

Likewise, no cleaning solutions should be introduced into the rainwater tank. If a system has a backup municipal water source, it is recommended that this water is not deposited into the tank as the treatment chemicals could kill the biofilm layer. The backup water should instead bypass the tank through solenoid valves and appropriate backflow prevention.

2.7 Pumps

Pumping can be accomplished in a couple of ways. Using a submersible cistern pump is one application and a booster/ suction pump is the other. Some applications will require a combination of pumps to meet the required pressures and gallons per minute necessary to meet the demands for the project.

2.7.1. Submersible Pumps

Well pumps can be used in rainwater systems, but if a well pump is used, it will need a constant flow of water passing by the motor to keep it cool. To accomplish this, the well pump can be placed in a cooling jacket which is attached to an intake device. Figure 2.15 shows a submersible pump draws water from the bottom of the pump. To prevent disturbing the biofilm, it is fitted with a floating suction intake.

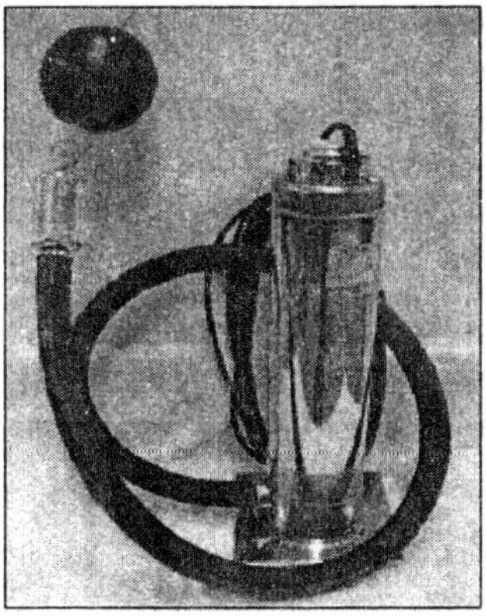

Figure 2.15. Submersible Pumps

2.7.2. Booster/Jet Pumps

External boester or suction pumps are popular in rainwater harvesting applications. These pumps rely on lifting the water out of the tank and pushing it to the desired location. External pumps are louder than submersible pumps and must be protected from the weather.

2.7.3. Pump Combinations

Some applications may require a submersible pump as well as a jet or booster pump. Normally this application is necessary when the distance from the cistern to the source is significant. The suction feature of the jet or booster pumps may not allow for the pump to keep its prime or supply enough water to meet the pumping requirements. Large applications may require two or more submersible pumps to meet the demand of the booster pump. When using this type of configuration, a controller for the pumps is necessary. Sizing these large pump systems should be accomplished by an experienced pump professional.

2.7.4. Pressure Tanks and Switches

Very few pumps have the ability to start on their own. It will be important to include a pressure switch and a pressure tank to a rainwater system requiring a pump. If the application uses more than one pump, the pressure switch will activate the control panel and the panel will determine which pump will run to meet the demand. It is also important to consider the size of the pressure tank to ensure that your system operates efficiently and effectively. The rule of thumb for sizing a pressure tank is three times the gallons per minute of the pump equals the gallon size of the pressure tank.

2.7.5. Pump Protection

Should your rainwater cistern become low on water, it will be essential that the pump is protected from running while there is not sufficient water for the pump to function. If the pump continues to run without sufficient water, the pump will be damaged. A simple float switch in the rainwater cistern will protect the pump from running when the water level declines. Some pumps include dry run protection; however, many of these pumps require manual resetting. A float switch automatically resets when the tank refills.

2.7.5.1. Floating Filters

To aspirate the water from the tank, a floating filter is located at the end of the pump's suction hose. As mentioned in the section on calming inlets, sediment, bacteria and other pollutants generally settle to the bottom of the tank, with concentrations of pollutants higher in the bottom sediments than at the surface of the water. Studies have also shown that bacteria levels can be higher at the surface of the cistern water than elsewhere in the tank. To avoid these two areas of concern, the floating filter takes water from the tank a few inches below the water surface. Floating filters should be designed to never clog, often by using a larger pore size than was used in the pre-tank filtration. Filter fabric should also be high-quality stainless steel, like the first flush filter.

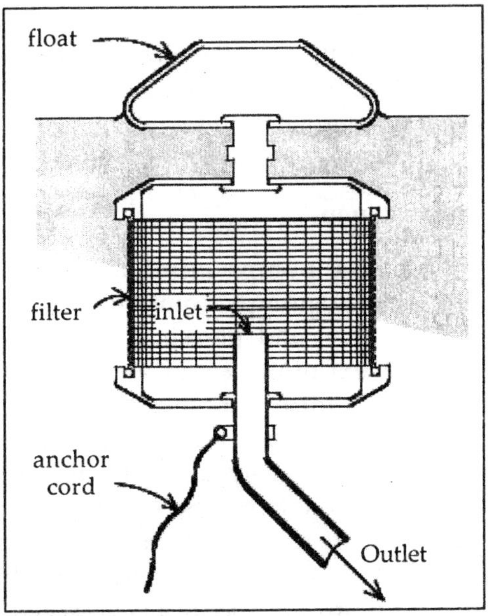

Figure 2.16. Cistern float filter

Pump systems draw water from the storage tanks, pressurise it, and store it in a pressure tank until needed. The typical pump-and-pressure tank arrangement consists of a ¾- or 1-horsepower pump, usually a shallow well jet pump or a multistage centrifugal pump, the check valve, and pressure switch. A one-way check valve between the storage tank and the pump prevents pressurised water from being returned to the tank. The pressure switch regulates operation of the pressure tank. The pressure tank, with a typical capacity of 40 gallons, maintains pressure throughout the system. When the pressure tank reaches a preset threshold, the pressure switch cuts off power to the pump. When there is demand from the household, the pressure switch detects the drop in pressure in the tank and activates the pump, drawing more water into the pressure tank.

The cistern float filter allows the pump to draw water from the storage tank from between 10 and 16 inches below the surface. Water at this level is cleaner and fresher than water closer to the bottom of the tank. The device has a 60-micron filter. An external suction pump, connected via a flexible hose, draws water through the filter.

2.7.6. On-demand Pump

The new on-demand pumps eliminate the need for a pressure tank. These pumps combine a pump, motor, controller, check valve, and pressure tank function all in one unit. They are self-priming and are built with a check valve incorporated into the suction port. A typical installation of an on-demand pump and a 5-micron fiber filter, 3-micron activated charcoal filter, and an ultraviolet lamp. Unlike conventional pumps, on-demand pumps are designed to activate in response to a demand, eliminating the need, cost, and space of a pressure tank. In addition, some on-demand pumps are specifically designed to be used with rainwater.

2.8. Overflow

All storage tanks must have an overflow of at least the same diameter as the inlet pipe to prevent backup in the gutters and downspouts or roof drains, but a properly designed overflow can also help protect the water quality in the tank. Overflows should be directed to a pervious area, ideally a stormwater management BMP. This practice will help reduce the chance that overflow from the rainwater harvesting system will generate a runoff problem. In these situations, backflow prevention is needed to prevent the introduction of small animals and prevent against the unlikely backup of water in the system. If discharge to a municipal storm system is the only option, backflow prevention should be increased because of the increased likelihood of a backup in the system. A properly designed overflow should also provide skimming of the water surface. Even with proper pre-tank filtration, some small buoyant debris, such as pollen, will likely enter the tank. A vertical pipe with a slight angled opening provides skimming of the water surface and removes this debris.

2.9. Treatment and Disinfection Equipment

The general public is often leery about consuming and utilising rainwater for potable and/or non-potable use, but proper system design makes harvested rainwater a safe water source. Additional treatment can be divided into three main categories: sediment filtration, disinfection and aesthetic treatme.

Disinfection of harvested rainwater for non-potable uses is sometimes seen as overly cautious. Because animals are the primary sources of bacteria to rainwater harvesting systems, bacteria found in rainwater systems may not be infectious to humans. The risk from non-potable uses is also low because rainwater should not be ingested in any non-potable use. However, Legionella, a bacterium that causes a type of pneumonia through inhalation, has been identified as a disease-causing agent in some rainwater harvesting systems. Legionella can be easily destroyed by disinfection techniques such as ultraviolet light, making disinfection of harvested rainwater for indoor non-potable uses advisable even though the disease risk is low.

2.9.1. Sediment Filtration

Sediment filtration removes fine particles which can carry contaminants and decrease the effectiveness of disinfection. While many of the potential contaminants found in rainwater harvesting systems can occur in both dissolved and solid forms, a high percentage is found bound to small particles. While many of these particles should have settled in the storage tank, the sediment filter should remove remaining small particles and associated contaminants. Small particles will also affect the clarity of the water and therefore the ability of ultraviolet light to kill bacteria and viruses.

2.9.2. Carbon Filtration

Activated carbon is particularly effective at removal of organic compounds, such as pesticides and some hydrocarbons. In addition to mechanical straining, activated carbon removes these contaminants because they adsorb, or stick, to the activated carbon. Activated carbon also frequently improves the aesthetic quality of water (i.e. odor and taste). An activated carbon filter of 3 micron or smaller should be included on all systems for potable use This filter should be placed after the sediment filter.

2.9.3. Disinfection Options

2.9.3.1. Chlorination

Chlorine reacts with water to form HOCl, which inactivates bacteria and viruses. The effectiveness of chlorine is dependent on the clarity of the water as well as the contact time with shorter contact times requiring higher doses of chlorine to achieve appropriate disinfection. Because the addition of chlorine to the storage tank would disrupt the biofilm, chlorine should not be added to the storage tank. Chlorine should instead be added using an injection system which may require higher concentrations of chlorine because of short contact times. One advantage of chlorine disinfection is the maintenance of a disinfectant residual in the system which should provide protection against the re-growth of bacteria. A chlorine residual of 0.2 ppm should be maintained in potable systems that use chlorine for disinfection.

2.9.3.2. *Ozonation*

The process by which ozone destroys bacteria and viruses is not well-understood, but ozone effectively destroys many proteins, which is likely part of the mechanism for ozone disinfection. While ozone is a highly effective disinfectant, it is generally less effective at destruction of viruses than destruction of bacteria or protozoa. Unlike chlorine, ozone does not leave a disinfectant residual in the finished water.

2.9.3.3. *Ultraviolet Light*

Ultraviolet light is a low-cost and effective method for treating bacteria and viruses. Ultraviolet light should only be used following a 5 micron sediment filter because ultraviolet light disinfection is only effective on very clear water. The appropriate dose of ultraviolet light for disinfection of harvested rainwater is 40,000 uw- sec/cm^2.

2.9.3.4. *Reverse Osmosis*

Reverse osmosis does not destroy or deactivate bacteria or viruses, but instead mechanically filters them out through a membrane. Reverse osmosis uses pressure to force water through a membrane from an area of high concentration of contaminants to an area of very low contaminants. The membranes typically exclude particles larger than about 0.0001 micron. Reverse osmosis is generally not a recommended disinfection method for rainwater harvesting because most reverse osmosis units waste large quantities of water.

2.9.4. Treatment for pH

The pH of rain and snow is typically acidic. This low pH level is sometimes cited as a concern in the use of rainwater harvesting systems. However, the pH of harvested rainwater is likely different from the pH of precipitation and most problems with pH can be easily corrected with low-cost, low-maintenance, but highly effective solutions. Multiple factors affect the pH of precipitation. Rain and snow are naturally slightly acidic because carbon dioxide in the atmosphere leads to development of a weak carbonic acid. However, fossil fuel use has led to emissions of sulfur dioxides and nitrogen oxides which further reduce the pH of rainfall. Acid rain is a widespread problem because atmospheric pollution is carried by the wind and spreads out far from the original source. A neutralising agent can be added to the tank to combat pH problems. Pieces of limestone rock can be placed in the tank to assist in neutralising the water. Other common neutralising agents and their dose per 1,000 gallons of water are as follows:

- Limestone: 2 oz.
- Quicklime: 1 oz.
- Hydrated lime: 1 oz.
- Soda ash: 1 oz.
- Caustic soda: 1.5 oz.

Another easy solution to problems of pH is to use a concrete storage tank. Limestone in the concrete will help neutralise the acid with no maintenance required.

The detailed description of water quality and treatment is decribed in Chapter Seven Water Quality and Treatment.

2.10. Back-up Water Supply and Plumbing

When utilising both rainwater and potable municipal water supplies, there may be times when rainwater supplies are exhausted, therefore municipal potable water or well water can be considered as a backup for non-potable water needs. To ensure cross contamination does not occur, separate plumbing/piping systems must be installed for both potable and non-potable rainwater. Signs should be displayed at all faucets and spigots supplying non-potable water that state not drinking water. Also, rainwater spigots must be protected, especially in public buildings, against unauthorised usage. Removing knobs from outside spigots is one approach to prevent unauthorised use. Potable and non-potable plumbing can be distinguished through different color pipes. Purple is often used to designate non-potable water piping. All non-potable water plumbing/ piping should be labeled as such. In the event of a rainwater shortage, tanks can be partially filled with potable municipal supplies or well water. However, filling the tank from a potable source is less desirable than simply having the end use of the rainwater be supplied by the alternate source during periods of low rainfall.

The introduction of treated, municipal water, which may contain chlorine, could have negative effects on the tank's biofilm. Further, it decreases the available volume of the tank for harvesting the next rain event. Rainwater harvesting systems should be designed to incorporate either a backflow prevention device or an air gap to ensure contamination does not occur between the rainwater and potable water supplies. The air gap should allow the space equivalent of 3 times the municipal water supply pipe diameter. The municipal water inlet must be installed above the highest possible rainwater level and overflow outlet to ensure cross contamination does not occur. Local building officials should be consulted to determine what type of backflow protection is desired in a jurisdiction.

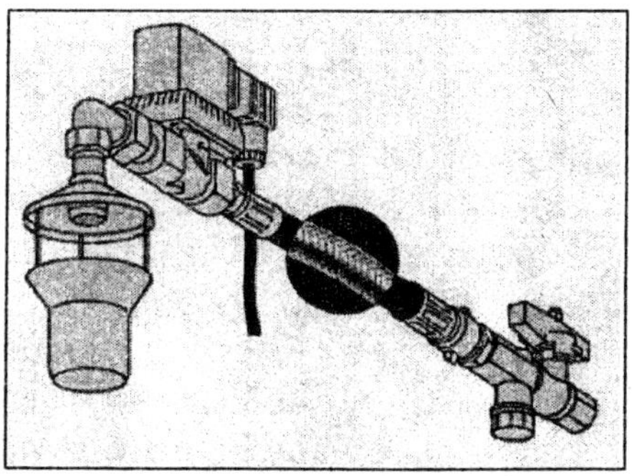

Figure 2.17. An air gap assembly

Backup water supplies can be activated by a variety of means. Commonly, if the water level in the rainwater harvesting tank becomes low, a float switch or switches will be activated shutting off the rainwater harvesting pump and activating a valve allowing water to flow from the backup source. An alternative is to use a pressure differential system. In this system the rainwater harvesting system operates at a higher pressure than the backup source. When the rainwater harvesting system pump cuts off due to a low water situation, the lower pressure water from the backup source is no longer held back by the higher pressure water and begins to flow to the fixtures being served by the rainwater harvesting system.

REFERENCES

Lye D. 2002. "Health risks associated with consumption of untreated water from household roof catchment systems". *Journal of the American Water Resources Association* 38(5):1301-1306.

Thomas PR, Grenne GR. "Rainwater quality from different roof catchments". *Water Science Technology* (28):290-99. 1993.

Texas Water Development Board. *Texas guide to rainwater harvesting.* Austin (TX): Texas Water Development Board. 58 p. 1997.

3 Traditional Methods of Rainwater Harvesting

Water has been harvested in India since antiquity, with our ancestors perfecting the art of water management. Many water harvesting structures and water conveyance systems specific to the eco-regions and culture has been developed. Our ancestors harvested the rain drop directly. From rooftops, they collected water and stored it in tanks built in their courtyards. From open community lands, they collected the rain and stored it in artificial wells.They harvested monsoon runoff by capturing water from swollen streams during the monsoon season and stored it various forms of water bodies. They harvested water from flooded rivers.

There are evidences that, even during Harappan period, there was very good system of water management as could be seen in the latest excavation at Dholavira in Kachch. During independence period, the people use to manage water resources considering it as part of the nature which is essential for their survival. This could be seen from the rainwater harvesting structures in the low rainfall areas of Rajasthan, harvesting springs in hilly areas and mountainous region and percolation ponds and tanks in southern India.

The ancient people stored rainwater in public placed separately one for drinking purposes and another for bathing and other domestic purpposses and called them as Ooranies. They also formed percolation tanks or ponds, for the purpose of recharging irrigation or domestic wells. They periodically clean the water ways so as to get clean water throughout the year. These are instances in the history that people constructed crude rubble bunas across river courses either for diversion of water or for augmenting the ground water. The various methods of rainwater harvesting are classified below under two category, Traditional and Modern methods.

Traditional rainwater harvesting, which is still prevalent in rural areas, was done in surface storage bodies like lakes, ponds, irrigation tanks, temple tanks etc. In urban areas, due to shrinking of open spaces, rainwater will have to necessarily be harvested as ground water, hence harvesting in such places will depend very much on the nature of the soil viz., clayey, sandy etc. This chapter descrioed various kinds of traditional rainwater harvesting methods.

3.1. Paar System

Paar is a common water harvesting practice in the western Rajasthan region. It is a common place where the rainwater flows from the agar (catchment) and in the process percolates into the sandy soil. In order to access the rajani pani (percolated water) kuis or beris are dug in the agor (storage area). Kuis or beris are normally 5 metres (m) to 12 m deep. The structure was constructed through traditional masonary technology.

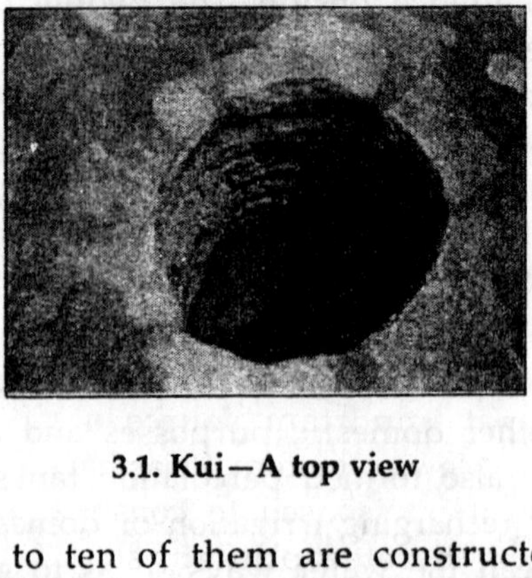

3.1. Kui—A top view

Normally six to ten of them are constructed in a paar. However depending on the size of the paar the numbers of kuis or beris are decided. Bhatti mentions that there are paars in Jaisalmer district where there are more than 20 kuis are in operation. This is the most predominant form of rainwater harvesting in the region. Rainwater harvested through PAAR technique is known as Patali paani.

3.2. Talab/Bandhis

Talabs are reservoirs. They may be natural, such as the ponds *(pokhariyan)* at Tikamgarh in the Bundelkhand region. They can be human-made, such the lakes in Udaipur. A reservoir area of less than five *bighas* is called a *talai;* a medium sized lake is called a *bandhi* or *talab;* bigger lakes are called *sagar* or *samand*. The *pokhariyan* serve irrigation and drinking purposes. When the water in these reserviors dries up just a few days after the monsoon, the pond beds are cultivated with rice.

3.2. A Talab

3.3. Saza Kuva

An open well with multiple owners (*saza* = partner), *saza kuva* is the most important source of irrigation in the Aravalli hills in Mewar, eastern Rajasthan. The soil dug out to make the well pit is used to construct a huge circular foundation or an elevated platform sloping away from the well. The first is built to accomodate the *rehat*, a traditional water lifting device; the sloping platform is for the *chada,* in which buffaloes are used to lift water.

3.3. *Saza kuva*

Saza kuva construction is generally taken up by a group of farmers with adjacent landholdings; a harva, a man with special skills in groundwater detection, helps fix the site.

3.4. JOHAD

Johads are small earthen check dams that capture and conserve rainwater, improving percolation and groundwater recharge.

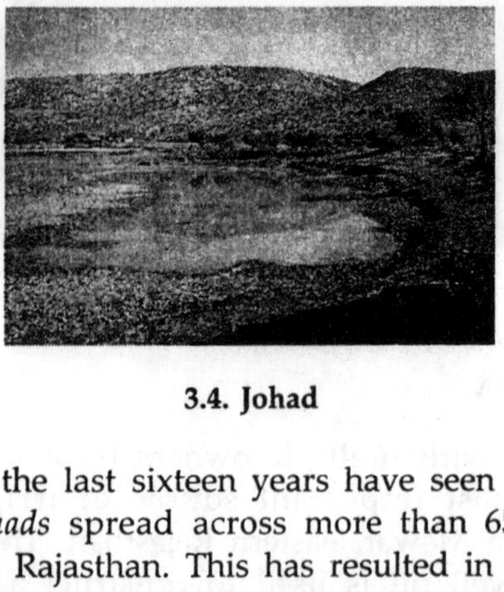

3.4. Johad

Starting 1984, the last sixteen years have seen the revival of some 3000 *johads* spread across more than 650 villages in Alwar district, Rajasthan. This has resulted in a general rise of the groundwater level by almost 6 metres and a 33 percent increase in the forest cover in the area. Five rivers that used to go dry immediately following the monsoon have now become perennial, such as the River Arvari, has come alive.

3.5. Pat System

Bhitada village, Jhabua district of Madhya pradesh developed the unique pat system. This system was devised according to the peculiarities of the terrain to divert water from swift-flowing hill streams into irrigation channels called pats.

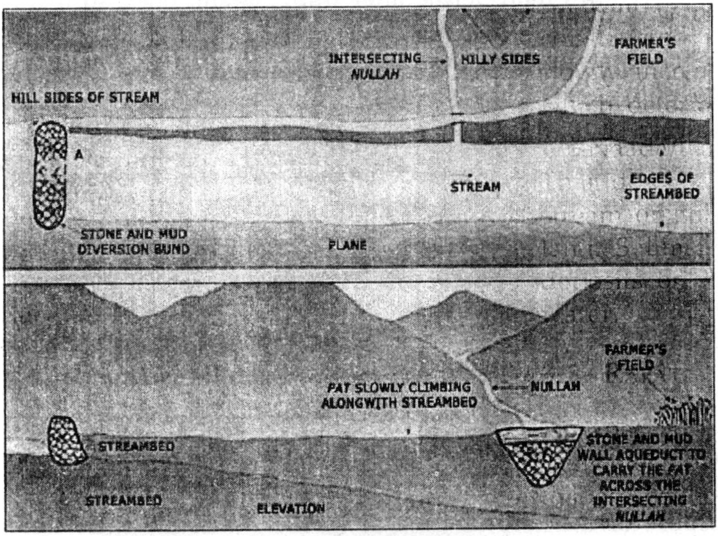

3.5. Pat system

The diversion bunds across the stream are made by piling up stones and then lining them with teak leaves and mud to make them leakproof. The pat channel has to negotiate small nullahs that join the stream off and on, and also sheer cliffs before reaching the fields. These sections invariably get washed away during the monsoons. Stone aqueducts have to be built to span the intervening nullahs.

The villagers irrigate their fields by turns. The channel requires constant maintenance and it is the duty of the family irrigating the fields on a particular day to take care of the pat on that particular day. It takes about two weeks to get the pat flowing and the winter crop is sown in early November.

3.6. Naada/Bandha

These are stone check dams across streams or gullies that are constructed to capture runoff on a stretch of fertile land that is submerged in water during the monsoon. The land not only becomes more fertile after trapping silt, but also retains substantial quantities of water in the soil. These dams are constructed in phases over several years. The height is slowly increased up to the right height of the check dam which determines the size of the naada.

Naada/bandha are found in the Mewar region of the Thar desert. It is a stone check dam, constructed across a stream or gully, to capture monsoon runoff on a stretch of land. Submerged in water, the land becomes fertile as silt deposits on it and the soil retains substantial amounts of water.

3.7. Rapat

A rapat is a percolation tank, with a bund to impound rainwater flowing through a watershed and a waste weir to dispose of the surplus flow. If the height of the structure is small, the bund may be built of masonary, otherwise earth is used. Rajasthan rapats, being small, are all masonry structures. Rapats and percolation tanks do not directly irrigate land, but recharges well within a distance of 3-5 km downstream. Silting is a serious problem with small rapats and the estimated life of a rapat varies from 5 to 20 years.

3.8. Chandela Tank

These tanks were constructed by stopping the flow of water in rivulets flowing between hills by erecting massive earthen embankments, having width of 60m or more. These hills with long stretches of quartz reefs running underneath them, acted as natural ground water barrier helping to trap water between the ridges. The earthen embankments were supported on both sides with walls of coarse stones, forming a series of stone steps. These tanks are made up of lime and mortar and this is the reason why these tanks survived even after thousand years but the only problem, which these tanks are facing, is siltation of tank beds. Chandela tanks usually had a convex curvature somewhere in the middle of the embankment; many older and smaller tanks were constructed near the human settlement or near the slopes of a cluster of hills. These tanks served to satisfy the drinking water needs of villagers and cattle.

3.9. Bundela Tank

These tanks are bigger in size as compared to Chandela tanks. These tanks had solidly constructed steps leading to water in the tank; But these structures had chabootaras, pavillions and royal orchards designed to show off the glory of the king who built them. But these tanks are not as cost effective and simple as Chandela tanks. These tanks were constructed to meet the growing water demands in the area, maintenance of these tanks was done by the person employed by the king but in case of smaller tanks villagers collectively removed silt and repair embankment.

3.10. Kunds/Kundis

A kund or kundi looks like an upturned cup nestling in a saucer. These structures harvest rainwater for drinking, and dot the sandier tracts of the Thar Desert in western Rajasthan and some areas in Gujarat. Essentially a circular underground well, *kunds* have a saucer-shaped catchment area that gently slopes towards the centre where the well is situated. A wire mesh across water-inlets prevents debris from falling into the well-pit. The sides of the well-pit are covered with (disinfectant) lime and ash. Most pits have a dome-shaped cover, or at least a lid, to protect the water. If need be, water can be drawn out with a bucket. The depth and diameter of *kunds* depend on their use (drinking, or domestic water requirements). They can be owned by only those with money to invest and land to construct it. Thus for the poor, large public *kunds* have to be built.

3.11. Beris

Found in western Rajasthan, these are 10-12 m deep pits dug near tanks to collect the seepage. *Kuis* can also be used to harvest rainwater in areas with meagre rainfall.The mouth of the pit is usually made very narrow. This prevents the collected water from evaporating. The pit gets wider as it burrows underunder the ground, so that water can seep in into a large surface area. The openings of these entirely *kuchcha* (earthen) structures are generally covered with planks of wood, or put under lock and key. The water is used sparingly, as a last resource in crisis situations. Magga Ram Suthar, of village Pithla in Jaisalmer district in Rajasthan, is an engineer skilled in making *kuis/beris*.

3.12. Baoris/Bers

Baoris or *bers* are community wells, found in Rajasthan, that are used mainly for drinking. Most of them are very old and were built by *banjaras* (mobile trading communities) for their drinking water needs. They can hold water for a long time because of almost negligible water evaporation.

3.13. Jhalaras

Jhalaras were human-made tanks, found in Rajasthan and Gujarat, essentially meant for community use and for religious rites. Often rectangular in design, *jhalaras* have steps on three or four sides. Jhalars areground water bodies which are built to ensure easy & regular supply of water to the surrounding areas, the jhalars are rectangular in shape with steps on three or even on all the four sides of the tank, the steps are built on a series of levels.The jhalaras collect subterranean seepage of a talab or a lake located upstream.The water from these jhalaras was not used for drinking but for only community bathing and religious rites.Jhodhpur city has eight jhalaras two of which are inside the town & six are found outside the city.The oldest jhalara is the mahamandir jhalara which dates back to 1660 AD.

3.14. Nadis

Nadis are village ponds, found near Jodhpur in Rajasthan. They are used for storing waterfrom an adjoining natural catchment during the rainy season. The site was selected by the villagers based on an available natural catchments and its water yield potential. Water availability from nadi would range from two months to a year after the rains. They are dune areas range from 1.5 to 4.0 metres and those in sandy plains varied from 3 to 12 metres. The location of the nadi had a strong bearing on its storage capacity due to the related catchment and runoff characteristics.

3.15. Tobas

Tobas is the local name given to a ground depression with a natural catchment area. A hard plot of land with low porosity, consisting of a depression and a natural catchment area was selected for the construction of tobas.

3.16. Tankas

Tankas (small tank) are underground tanks, found traditionally in most Bikaner houses. They are built in the main house or in the courtyard. They were circular holes made in the ground, lined with fine polished lime, in which raiwater was collected. *Tankas* were often beautifully decorated with tiles, which helped to keep the water cool. The water was used only for drinking. If in any year there was less than normal rainfall and the *tankas* did not get filled, water from nearby wells and tanks would be obtained to fill the household *tankas*. In this way, the people of Bikaner were able to meet their water requirements. The *tanka* system is also to be found in the pilgrim town of Dwarka where it has been in existence for centuries. It continues to be used in residential areas, temples, *dharamshalas* and hotels.

3.17. KHADIN

A khadin, also called a *dhora,* is an ingenious construction designed to harvest surface runoff water for agriculture. Its main feature is a very long (100-300 m) earthen embankment built across the lower hill slopes lying below gravelly uplands. Sluices and spillways allow excess water to drain off. The *khadin* system is based on the principle of harvesting rainwater on farmland and subsequent use of this water-saturated land for crop production.

3.6. Khadin

First designed by the Paliwal Brahmins of Jaisalmer, western Rajasthan in the 15th century, this system has great similarity with the irrigation methods of the people of Ur (present Iraq) around 4500 BC and later of the Nabateans in the Middle East. A similar system is also reported to have been practised 4,000 years ago in the Negev desert, and in southwestern Colorado 500 years ago.

3.18. VAV/VAVDI/BAOLI/BAVADI

Traditional stepwells are called *vav* or *vavadi* in Gujarat, or *baolis* or *bavadis* in Rajasthan and northern India. Built by the nobility usually for strategic and/or philanthropical reasons, they were secular structures from which everyone could draw water. Most of them are defunct today.

The construction of stepwells date from four periods: Pre-Solanki period (8th to 11th century CE); Solanki period (11th to 12th century CE); Vaghela period (mid-13th to end-14th century CE); and the Sultanate period (mid-13th to end-15th century CE).Sculptures and inscriptions in stepwells demonstrate their importance to the traditional social and cultural lives of people.

Stepwell locations often suggested the way in which they would be used. When a stepwell was located within or at the edge of a village, it was mainly used for utilitarian purposes and as a cool place for social gatherings. When stepwells were located outside the village, on trade routes, they were often frequented as resting places. Many important stepwells are located on the major military and trade routes from Patan in the north to the sea coast of Saurashtra. When stepwells were used exclusively for irrigation, a sluice was constructed at the rim to receive the lifted water and lead it to a trough or pond, from where it ran through a drainage system and was channelled into the fields.A major reason for the breakdown of this traditional system is the pressure of centralisation and agricultural intensification.

3.19. Ahar Pynes

This traditional floodwater harvesting system is indigenous to south Bihar.In south Bihar, the terrain has a marked slope — 1 m per km — from south to north. The soil here is sandy and does not retain water. Groundwater levels are low. Rivers in this region swell only during the monsoon, but the water is swiftly carried away or percolates down into the sand. All these factors make floodwater harvesting the best option here, to which this system is admirably suited. An *ahar* is a catchment basin embanked on three sides, the 'fourth' side being the natural gradient of the land itself. Ahar beds were also used to grow a *rabi* (winter) crop after draining out the excess water that remained after *kharif* (summer) cultivation.

Pynes are artificial channels constructed to utilise river water in agricultural fields. Starting out from the river, *pynes* meander through fields to end up in an ahar. Most *pynes* flow within 10 km of a river and their length is not more than 20 km.The ahar-pyne system received a death-blow under the nineteenth-century British colonial regime. The post-independent state was hardly better. In 1949, a Flood Advisory Committee investigating continuous floods in Bihar's Gaya district came to the conclusion that "the fundamental reason for recurrence of floods was the destruction of the old irrigational system in the district."

Of late, though, some villages in Bihar have taken up the initiative to re-build and re-use the system. One such village is Dihra.It is a small village 28 km southwest of Patna city. In 1995, some village youths realised that they could impound the waters of the Pachuhuan (a seasonal stream passing through the village that falls into the nearby river Punpun) and use its bed as a reservoir to meet the village's irrigation needs. Essentially, this meant creating an ahar-pyne systemAfter many doubts, the village powers-that-be gave the go-ahead. Money was collected and work began in May 1995. After a month of shramdaan (voluntary labour) the villagers completed their work mid-June.Their efforts have borne fruit. By 2000 AD, the ahar was irrigating 80 ha of land. The people grow two cereal crops and one crop of vegetables every year. The returns from the sale of what they produce are good. The village is no longer a poor one.

3.20. Inundation Channel

Bengal once had an extraordinary system of inundation canals. Sir William Willcocks, a British irrigation expert who had also worked in Egypt and Iraq, claimed that inundation canals were in vogue in the region till about two centuries ago. Floodwater entered the fields through the inundation canals, carrying not only rich silt but also fish, which swam through these canals into the lakes and tanks to feed on the larva of mosquitoes. This helped to check malaria in this region. According to Willcocks, the ancient system of overflow irrigation had lasted for thousands of years. Unfortunately, during the Afghan-Maratha war in the 18th century and the subsequent British conquest of India, this irrigation system was neglected, and was never revived. The distinguishing features of the irrigation system were:

1. the canals were broad and shallow, carrying the crest waters of the river floods, rich in fine clay and free from coarse sand;
2. the canals were long and continuous and fairly parallel to each other, and at the right distance from each other for purposes of irrigation;
3. irrigation was performed by cuts in the banks of the canals, which were closed when the flood was over.

3.21. Dungs or Jampois

Dungs or Jampois are small irrigation channels linking rice fields to streams in the Jalpaiguri district of West Bengal.

3.22. Cheruvu

Cheruvu are found in Chitoor and Cuddapah districts in Andhra Pradesh. They are reservoirs to store runoff. *Cheruvu* embankments are fitted with *thoomu* (sluices), *alugu* or *marva* or *kalju* (flood weir) and *kalava* (canal).

3.23. Kohli Tanks

The Kohlis, a small group of cultivators, built some 43,381 water tanks in the district of Bhandara, Maharashtra, some 250-300 years ago. These tanks constituted the backbone of irrigation in the area until the government took them over in the 1950s. It is still crucial for sugar and rice irrigation. The tanks were of all sizes, often with provisions to bring water literally to the doorstep of villagers.

3.24. Bhanadaras

These are check dams or diversion weirs built across rivers.

A traditional system found in Maharashtra, their presence raises the water level of the rivers so that it begins to flow into channels. They are also used to impound water and form a large reservoir.Where a *bandhara* was built across a small stream, the water supply would usually last for a few months after the rains. They are built either by villagers or by private persons who received rent-free land in return for their public actMost Bandharas are defunct today. A very few are still in use.

3.25. Phad

The community-managed phad irrigation system, prevalent in northwestern Maharashtra, probably came into existence some 300-400 years ago. The system operated on three rivers in the Tapi basin—Panjhra, Mosam and Aram—in Dhule and Nasik districts.

The system starts with a *bandhara* (check dam or diversion-weir) built across a rivers. From the *bandharas* branch out *kalvas* (canals) to carry water into the fields. The length of these canals varies from 2-12 km. Each canal has a uniform discharge capacity of about 450 litres/second. *Charis* (distributaries) are built for feeding water from the kalva to different areas of the *phad. Sarangs* (field channels) carry water to individual fields. *Sandams* (escapes), along with *kalvas* and *charis,* drain away excess water. In this way water reaches the *kayam baghayat* (agricultural command area), usually divided into four *phads* (blocks).

The size of a *phad* can vary from 10-200 ha, the average being 100-125 ha. Every year, the village decides which phads to use and which to leave fallow. Only one type of crop is allowed in one phad. Generally, sugarcane is grown in one or two phads; seasonal crops are grown in the others. This ensures a healthy crop rotation system that maintains soil fertility, and reduces the danger of waterlogging and salinity. The *phad* system has given rise to a unique social system to manage water use.

3.26. KERE

Tanks, called kere in Kannada, were the predominant traditional method of irrigation in the Central Karnataka Plateau, and were fed either by channels branching off from *anicuts* (chech dams) built across streams, or by streams in valleys. The outflow of one tank supplied the next all the way down the course of the stream; the tanks were built in a series, usually situated a few kilometres apart. This ensured a) no wastage through overflow, and b) the seepage of a tank higher up in the series would be collected in the next lower one.

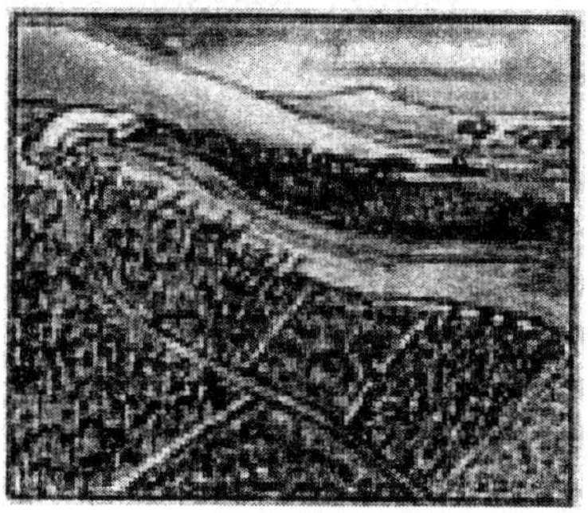

3.7. Kere

The Ramtek model has been named after water harvesting structures in the town of Ramtek, Maharashtra. A scientific analysis revealed an intricate network of groundwater and surface waterbodies, intrinsically connected through surface and underground canals. A fully evolved system, this model harvested runoff through tanks, supported by high yielding wells and structures like baories, kundis, and waterholes. This system, intelligently designed to utlise every raindrop falling in the watershed area is disintegrating due to neglect and ignorance.

Constructed and maintained mostly by *malguzars* (landowners), these tanks form a chain, extending from the foothills to the plains, conserving about 60-70 per cent of the total runoff. Once tanks located in the upper reaches close to the hills were filled to capacity, the water flowed down to fill successive tanks, generally through interconnecting channels. This sequential arrangement generally ended in a small waterhole to store whatever water remained unstored.

The presence of the Ramtek ridge in the middle, having a steep slope on both sides, results in quick runoffs and little percolation. This might have led the residents of the southern plains of the Ramtek hills to construct different types of water conservation structures (like tanks) where they could trap the maximum

3.27. Zings

Zings are water harvesting structures found in Ladakh. They are small tanks, in which collects melted glacier water. Essential to the system is the network of guiding channels that brings the water from the glacier to the tank. As glaciers melt during the day, the channels fill up with a trickle that in the afternoon turns into flowing water. The water collects towards the evening, and is used the next day.A water official called the *churpun* ensures that water is equitably distributed.

3.28. Kul

Kuls are water channels found in precipitous mountain areas. These channels carry water from glaciers to villages in the Spiti valley of Himachal Pradesh. Where the terrain is muddy, the *kul* is lined with rocks to keep it from becoming clogged. In the Jammu region too, similar irrigation systems called *kuhls* are found.

3.29. Naula

Naula is a surface-water harvesting method typical to the hill areas of Uttaranchal. These are small wells or ponds in which water is collected by making a stone wall across a stream.

3.8. Naula

3.30. Khatri

Khatris are structures, about 10x12 feet in size and six feet deep carved out in the hard rock mountain. The specially trained masons construct them at a cost of Rs 10,000-20,000 each. These traditional water harvesting structures are found in Hamirpur, Kangra and Mandi districts of Himachal Pradesh. There are two types of khatris: one for animals and washing purposes in which rainwater is collected from the roof through pipes, and other used for human consumption in which rainwater is collected by seepage through rocks. Interestingly, the khatris are owned by individual as well as by a community. There are government khatris as well, which are maintained by the panchayat.

3.31. Kuhl

Kuhls are a traditional irrigation system in Himachal Pradesh—surface channels diverting water from natural flowing streams (*khuds*). A typical community *kuhl* services six to 30 farmers, irrigating an area of about 20 ha. The system consists of a temporary headwall (constructed usually with river boulders) across a *khud* (ravine) for storage and diversion of the flow through a canal to the fields. By modern standards, building *kuhls* was simple, with boulders and labour forming the major input. The *kuhl* was provided with *moghas* (*kuchcha* outlets) to draw out water and irrigate nearby terraced fields. The water would flow from field to field and surplus water, if and, would drain back to the *khud*.

The kuhls were constructed and maintained by the village community. At the beginning of the irrigation season, the kohli (the water tender) would organise the irrigators to construct the headwall, repair the kuhl and make the system operational. The kohli played the role of a local engineer. Any person refusing to participate in construction and repair activities without valid reason, would be denied water for that season. Since denial of water was a religious punishment, it ensured community participation and solidarity. A person was also free to participate by providing a substitute for his labour. The kohli also distributed and managed the water.

3.32. Zabo

The *zabo* (the word means 'impounding run-off') system is practiced in Nagaland in north-eastern India. Also known as the *ruza* system, it combines water conservation with forestry, agriculture and animal care.Villages such as Kikruma, where zabos are found even today, are located on a high ridge. Though drinking water is a major problem, the area receives high rainfall. The rain falls on a patch of protected forest on the hilltop; as the water runs off along the slope, it passes through various terraces. The water is collected in pond-like structures in the middle terraces; below are cattle yards, and towards the foot of the hill are paddy fields, where the run-off ultimately meanders into.

3.33. Cheo-ozihi

The river Mezii flows along the Angami village of Kwigema in Nagaland. The riverwater is brought down by a long channel. From this channel, many branch channels are taken off, and water is often diverted to the terraces through bamboo pipes. One of the channels is named *Cheo-oziihi* - *oziihi* means water and *Cheo* was the person responsible for the laying of this 8-10 km-long channel with its numerous branches. This channel irrigates a large number of terraces in Kwigwema, and some terraces in the neighbouring village. There are three *khels* and the village water budget is divided among them.

3.34. Eri

Approximately one-third of the irrigated area of Tamil Nadu is watered by *eris* (tanks). *Eris* have played several important roles in maintaining ecological harmony as flood-control systems, preventing soil erosion and wastage of runoff during periods of heavy rainfall, and recharging the groundwater in the surrounding areas. The presence of *eris* provided an appropriate micro-climate for the local areas. Without *eris*, paddy cultivation would have been impossible.Till the British arrived, local communities maintained *eris*.

3.9. Eri

Historical data from Chengalpattu district, for instance, indicates that in the 18th century about 4-5 per cent of the gross produce of each village was allocated to maintain *eris* and other irrigation structures. Assignments of revenue-free lands, called *manyams*, were made to support village functionaries who undertook to maintain and manage *eris*. These allocations ensured *eri* upkeep through regular desilting and maintenance of sluices, inlets and irrigation channels.

The early British rule saw disastrous experiments with the land tenure system in quest for larger land revenues. The enormous expropriation of village resources by the state led to the disintegration of the traditional society, its economy and polity. Allocations for maintenance of *eris* could no longer be supported by the village communities, and these extraordinary water harvesting systems began to decline.Read more about Ganesan, the neerkatti who managed *eris*

3.35. Ooranis

The tanks, in south Travancore, though numerous, were in most cases oornis containing just enough water to cultivate the few acres of land dependent on them. The irregular topography of the region and the absence of large open spaces facilitated the construction of only small tanks unlike large ones seen in the flat districts of the then Madras Presidency, now Tamil Nadu.

3.36. Dongs

Dongs are ponds constructed by the Bodo tribes of Assam to harvest water for irrigation. These ponds are individually owned with no community involvement.

3.37. Bamboo Drip Irrigation

Meghalaya has an ingenious system of tapping of stream and springwater by using bamboo pipes to irrigate plantations. About 18-20 litres of water entering the bamboo pipe system per minute gets transported over several hundred metres and finally gets reduced to 20-80 drops per minute at the site of the plant. This 200-year-old system is used by the tribal farmers of Khasi and Jaintia hills to drip-irrigate their black pepper cultivation.Bamboo pipes are used to divert perennial springs on the hilltops to the lower reaches by gravity. The channel sections, made of bamboo, divert and convey water to the plot site where it is distributed without leakage into branches, again made and laid out with different forms of bamboo pipes. Manipulating the intake pipe positions also controls the flow of water into the lateral pipes.

Reduced channel sections and diversion units are used at the last stage of water application. The last channel section enables the water to be dropped near the roots of the plant. Bamboos of varying diameters are used for laying the channels. About a third of the outer casing in length and internodes of bamboo pieces have to be removed while fabricating the system. Later, the bamboo channel is smoothened by using a dao, a type of local axe, a round chisel fitted with a long handle. Other components are small pipes and channels of varying sizes used for diversion and distribution of water from the main channel. About four to five stages of distribution are involved from the point of the water diversion to the application point.

3.38. Apatani

This is a wet rice cultivation cum fish farming system practiced in elevated regions of about 1600 m and gentle sloping valleys, having an average annual rainfall about 1700 mm and also rich water resources like springs and streams. This system harvests both ground and surface water for irrigation. It is practiced by Apatani tribes of ziro in the lower Subansiri district of Arunachal Pradesh.

In Apatani system, valleys are terraced into plots separated by 0.6 meters high earthen dams supported by bamboo frames. All plots have inlet and outlet on opposite sides. The inlet of lowlying plot functions as an outlet of the high lying plot. Deeper channels connect the inlet point to outlet point. The terraced plot can be flooded or drained off with water by opening and blocking the inlets and outlets as and when required. The stream water is tapped by constructing a wall of 2-4 m high and 1 m thick near forested hill slopes. This is conveyed to agricultural fields through a channel network.

3.39. Virdas

Virdas are shallow wells dug in low depressions called *jheels* (tanks). They are found all over the Banni grasslands, a part of the Great Rann of Kutch in Gujarat. They are systems built by the nomadic Maldharis, who used to roam these grasslands. Now settled, they persist in using *virdas*. These structures harvest rainwater. The topography of the area is undulating, with depressions on the ground. By studying the flow of water during the monsoon, the Maldharis identify these depressions and make their *virdas* there.

3.10. Virda

Essentially, the structures use a technology that helps the Maldharis separate potable freshwater from unpotable salt water. After rainwater infiltrates the soil, it gets stored at a level above the salty groundwater because of the difference in their density. A structure is built to reach down (about 1 m) to this upper layer of accumulated rainwater. Between these two layers of sweet and saline water, there exists a zone of brackish water. As freshwater is removed, the brackish water moves upwards, and accumulates towards the bottom of the *virda*.

3.39. Katas/Mundas/Bandhas

The katas, mundas and bandhas were the main irrigation sources in the ancient tribal kingdom of the Gonds (now in Orissa and Madhya Pradesh). Most of these katas were built by the village headmen known as gountias, who in turn, received the land from the Gond kings. Land here is classified into four groups on the basis of its topography: *aat*, (highland); *mal* (sloped land); *berna* (medium land); and *bahal* (low land). This classification helps to select A *kata* is constructed north to south, or east to west, of a village.

A strong earthen embankment, curved at either end, is built across a drainage line to hold up an irregularly-shaped sheet of water. The undulations of the country usually determine its shape as that of a long isosceles triangle, of which the dam forms the base. It commands a valley, the bottom of which is the *bahal* land and the sides are the *mal* terrace. As a rule, there is a cut high up on the slope near one end of the embankment from where water is led either by a small channel or tal, or from field to field along terraces, going lower down to the fields. In years of normal rainfall, irrigation was not needed because of moisture from percolation and, in that case, the surplus flow was passed into a nullah. In years of scanty rainfall, the centre of the tank was sometimes cut so that the lowest land could be irrigated.

3.40. Surangam

Kasaragod district in the northern Malabar region of Kerala is an area whose people cannot depend directly on surface water. The terrain is such that there is high discharge in rivers in the monsoon and low discharge in the dry months. People here depend, therefore on groundwater, and on a special water harvesting structure called *surangam*.The word *surangam* is derived from a Kannada word for tunnel. It is also known as *thurangam, thorapu, mala,* etc, in different parts of Kasaragod. It is a horizontal well mostly excavated in hard laterite rock formations.

The excavation continues until a good amount of water is struck. Water seeps out of the hard rock and flows out of the tunnel. This water is usually collected in an open pit constructed outside the *surangam*. A *surangam* is about 0.45-0.70 metres (m) wide and about 1.8-2.0 m high. The length varies from 3-300 m. Usually several subsidiary *surangams* are excavated inside the main one. If the *surangam* is very long, a number of vertical air shafts are provided to ensure atmospheric pressure inside. The distance between successive air shafts varies between 50-60 m.

The approximate dimensions of the air shafts are 2 m by 2 m, and the depth varies from place to place. *Surangams* are similar to qanats which once existed in Mesopotamia and Babylon around 700 BC.1,2 By 714 BC, this technology had spread to Egypt, Persia (now Iran) and India. The initial cost of digging a *surangam* (Rs 100-150 per 0.72 m dug) is the only expenditure needed, as it hardly requires any maintenance. Traditionally, a *surangam* was excavated at a very slow pace and was completed over generations.

3.41. Korambus

Korambu is a temporary dam stretching across the mouth of channels, made of brushwood, mud and grass. It is constructed by horizontally fixing a strong wooden beam touching either banks of the canal. A series of vertical wooden beams of appropriate height is erected with their lower ends resting firmly on the ground and the other ends tied to the horizontal beam. Closely knitted or matted coconut thatch is tied to this frame. A coat of mud is applied to the matted frame.

A layer of grass is also applied carefully which prevents dissolution of the applied mud. Korambu is constructed to raise the water level in the canal and to divert the water into field channels. It is so built that excess water flows over it and only the required amount of water flows into the diversion channels. The height of the Korambu is so adjusted that the fields lying on the upstream are not submerged. Water is allowed to flow from one field to another until all the field are irrigated. They are built twice a year especially before the onset of the monsoon season in order to supply water during winter and summer season. In Kasargod and Thrissur districts of Kerala, Korambu is known as chira.

3.42. Jackwells

The difference in the physiography, topography, rock types and rainfall meant that the tribes in the different islands followed different methods of harvesting rain and groundwater. For instance, the southern part of the Great Nicobar Island near Shastri Nagar has a relatively rugged topography in comparison to the northern part of the islands. The shompen tribals here made full use of the topography to harvest water.

In lower parts of the undulating terrain, bunds were made using logs of hard bullet wood, and water would collect in the pits so formed. They make extensive use of split bamboos in their water harvesting systems. A full length of bamboo is cut longitudinally and placed along a gentle slope with the lower end leading into a shallow pit. These serve as conduits for rainwater which is collected drop by drop in pits called Jackwells. Often, these split bamboos are placed under trees to harvest the throughfalls (of rain) through the leaves. A series of increasingly bigger jackwells is built, connected by split bamboos so that overflows from one lead to the other, ultimately leading to the biggest jackwell, with an approximate diameter of 6 m and depth of 7 m so that overflows from one lead to the other.

References

Agarwal, A.; Narain, S. *Dying Wisdom: Rise and Fall of Traditional Water Harvesting Systems,* Centre for Science and Environment. New Delhi, India: Centre for Science and Environment. 1997.

Arya, S.L., Samra, J.S. 2001. *Revisiting watershed management institutions in Haryana Shivaliks, India.* Central soil and Water

Conservation Research and Training Institute, Chandigarh, India, pp. 329.

Gould, J., and E. Nissen-Petersen. *Rainwater Catchment Systems for Domestic Supply: Design, Construction and Implementation.* London, U.K.: IT Publications, 1999.

Pacey A., and A. Cullis. *Rainwater Harvesting: The Collection of Rainfall and Runoff in Rural Areas.* London, U.K.: IT Publications, 1986.

Sahuquillo, A. Groundwater in Water Resource Planning: Conjunctive Use, *Water International,* 10. 57-63. 1985.

Sharma, B.R, et.al., (eds.). *Watershed Management Challenges: Improving productivity, resources and livelihoods.* International Water Management Institute, Colombo, Sri Lanka. 2005.

4 | Modern Methods of Rainwater Harvesting

The numbers of water stressed regions at various part of the world are increasing due to the rapid growth rate of real estate. The urban areas are facing twin challenges of water scarcity and inadequate capacity of wastewater disposal systems. The rapid growth of urban population leads to escalation of water demand. Meeting these increasing water demand with a geographically constrained water supply system is often a very difficult task. Large construction projects are also being taken up at areas where municipal water supply is yet to be made available. The developers are frequently exploring and abstracting groundwater to ensure the basic amenities at their housing projects. Conservation of ground water is important because it takes years to be replenished. In areas where ground water is used, care must be taken to minimise the quantity of water withdrawn and bring it on per with quantity of water being replenished.

To address the challenges of water security in the new millenium, a mixture of traditional wisdom and new techniques must be employed. The modern methods of rainwater harvesting can be broadly categorised under two:

1. Collection and storage of rainwater for direct use, and
2. Groundwater recharging.

A combination of these two are also practiced, where rainwater is collected and stored in containers for direct use while the collected rainwater in excess of the storage capacity is charged into ground (groundwater recharging).

The underlying principle of the rainwater harvesting is to "collect the rainwater when it rains and don't allow rainwater to run away" (runoff). It requires arrangements for collection (usually through gutters and drop down pipelines) of rainwater falling on rooftop and storage of collected rainwater for direct use and/or for groundwater recharge. This method is commonly known as Rooftop Rainwater Harvesting. Rainwater falling on the open spaces around the building (other than rooftop) is also used for rainwater harvesting generally by means of groundwater recharging.

In urban areas, rainwater available from rooftop of buildings, paved areas are drained out. This water can be stored or recharged to aquifer and can be utilised gainfully at the time of need. In urban area, rainwater-harvesting system are designed in a way that it does not occupy large space for collection and recharge system. A few of the techniques of roof top rainwater harvesting in urban areas for recharging of groundwater aquifer are:

(a) roof top rainwater harvesting through recharge pit,

(b) roof top rainwater harvesting through recharge trench,

(c) roof top rainwater harvesting through existing tube wells and

(d) roof top rainwater harvesting through trench with recharge well.

Rainwater collected from terraces of a cluster of building may be led into nearby ponds (with previous top layer) or in a storage tank. Runoff water can also be diverted into such ponds.

The water shall be recommended for utilisation in car washing, landscaping, fire fighting etc. Different percentage of storage and recharge has been proposed for different stories of building to keep the cost escalation rate as well as rainwater harvested equitable irrespective of buildings heights.

Some of the modern systems that are being used for raiwater harvesting are described in this chapter The modern methods of rainwater harvesting are categorised under two. They are artifical recharging and rainwater harvesting. The former is classified into absorption pit method, absorption well method, well cum bore method and recharge trench cum injection well. The later is categorised into individual houses and grouped houses which are further classified into percolation pit method, bore well with settlement tank, open well method with filter bed pump and percolation pit with bore method.

4.1. Artificial Recharging

4.1.1. Absorption Pit Method

Apercolation/absorption pit is a hand bore made in the soil with the help of an augur and filled up with pebbles and river sand on top. The depth of these pits will be anywhere between 4 and 8 meters depending on the nature of the soil. If the soil is clayey, the pit has to be dug to a depth till a reasonably sandy stratum is reached. The diameter of these pits will be 25 cm (10 inches). A square / circular collection chamber with silt arrester is provided at the top.

The pit is constructed in the open space at required intervals.

— Size 1m x 1m x 1.5m (depth)

— Filled with broken bricks / pebbles

— Suitable for sandy sub - soil area

— One unit for 300 sq.ft area (approx.)

— A borehole to be drilled at the bottom of the percolation pit. Bore hole size : 150 - 300 mm dia with 10 -15 ft depth. Filled with broken bricks. Suitable for clay area.

An absorption pit should be constructed in the sloped area of the house veranda. The size of the absorption pit should be a minimum of 5 feet in width and a maximum of 10 feet in depth. Matching cement rings should be installed till the base of the pit. The pit should be filled with boulders (or big jelly stones) and a nylon curtain should be laid at the top. The remaining part of the pit should be filled with sand till one foot of the ground level.

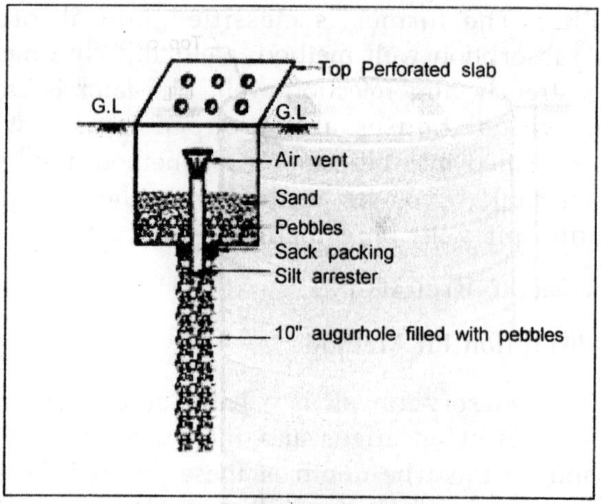

4.1. Percolation pit

The pipe, which is installed for collecting the rainwater from the rooftop and the veranda of the house, should be linked to this absorption pit. Another pipe should be deployed for allowing the excess water to run off into the storm water drainage. Thus, by effectively preventing the rainwater from running off into the drainage and by making it absorbed into the ground, ground water sources can be recharged. All Bangaloreans, by adopting the ground water recharging system, can save the future generations the problem of water shortages.

4.1.2. Absorption Well Method

These wells are constructed using cement rings readily available in the market. The diameter of these rings range from 2 ft to 6 ft. The depth to which these wells are dug depends on the nature of the soil and the diameter depends on the number of roof top pipes that are likely to be connected to each one of these wells. These wells are left unfilled and are covered with RCC slabs of suitable thickness to facilitate free pedestrian and vehicular movement on the ground.

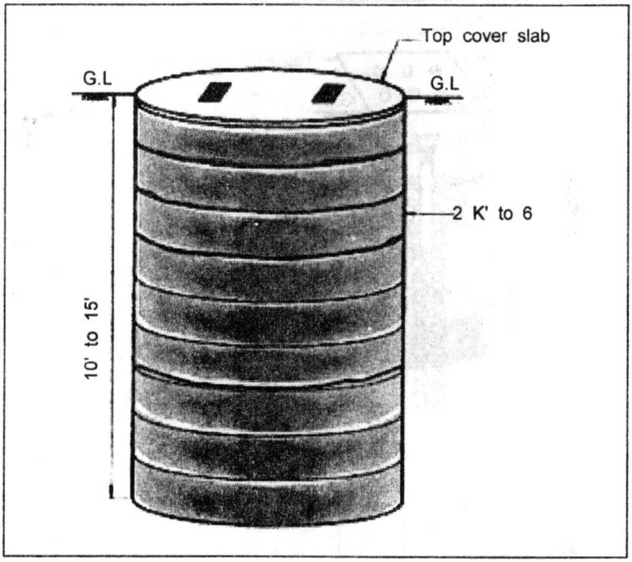

4.2. Percolation well

Rainwater from the terrace is diverted to the existing open well using PVC pipes through a filter chamber.

The minimum size of the filter chamber is 2" x 2" x 2" filled with broken bricks in the bottom and sand on the top. The chamber may be covered with RCC slab.

4.1.3. Well Cum Bore Method

In areas where the soil is likely to be clayey upto say 15ft. and more, it is advisable to go in for a percolation well upto 10ft. or 15ft. and a hand bore pit within this well upto a depth of 10ft. to 15ft. from its bottom.

A PVC pipe of 6in. diameter is inserted into the bore for the entire length.

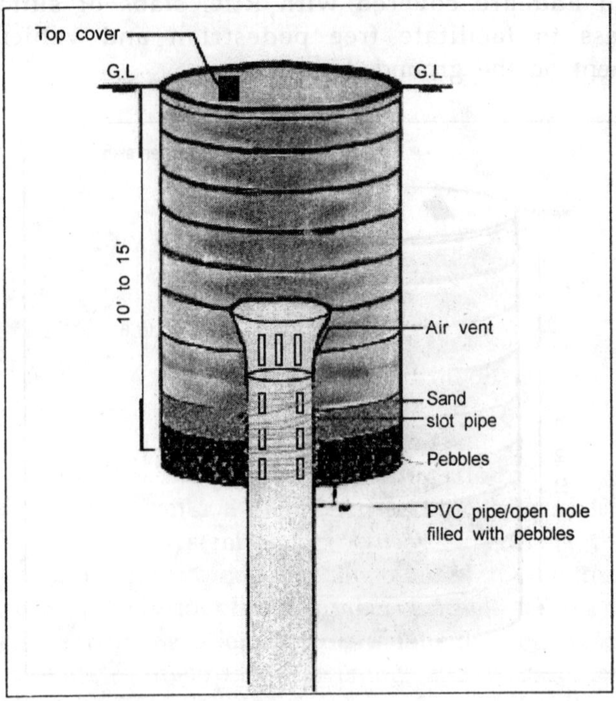

4.3. Percolation well cum borepit (for clayey soil)

4.1.4. Recharge Trench Cum Injection Well

This technique is ideally suited for areas where permeable sandy horizons within 3 to 5 m. below ground level and continues up to the water level under unconfined conditions, by which copious water available can be easily recharged.

In this technique, 1 to 2 m. wide and 2 to 3 m. deep trench is excavated, the length of which depends on the site availability and volume of water to be handled. An injection well of 100 to 150 mm dia meter is constructed, piereing through the layers of impermeable horizons to the potential aquifer reaching about 3 to 5 metres below water levels from the bottom of the trenches. Depending upon the volume of water to be injected, the number of injection wells can be increased to enhance the recharging rate. A schematic diagram is enclosed which is self explanatory.

4.1.4. Borewell Recharging

Recharging of borewells should be done to prevent them from drying up and improve their water table. It is usually done by the following method. First, a pit should be dug in the region surrounding the casing pipe and cement rings should be installed in it. The size of the pit should be one meter in diameter and 10 feet in depth. At the bottom of the pit, filter holes should be made and a casing pipe with steel mesh should be fixed tightly to the bore well pipe. This casing pipe will function as a filter. Cement rings matching the borewell's diameter should be installed till the base of the pit. The pit should be filled with boulders from the base up to a height of two feet. A second layer of two feet height filled with jelly stones of 40 mm should follow this. The third layer of one-foot height should be filled with jelly stones of 20 mm size. The fourth layer should be filled with charcoal. This exercise should be repeated till there is 3 feet of space left from form the ground level. A nylon curtain should be spread on the layers and the remaining space of the pit should be filled with sand until one foot from the ground level. A pipe should be fixed to collect the rainwater from the roof and this pipe should be connected to the pit. A pipe should be fixed to prevent the excess water from running away to roadside drainage.

4.1.6. Ground Water Recharging

Ground water recharging in urban areas is done by collecting the rainwater from the rooftops and the portico of the house and by making it easily absorbed within the veranda space. The method used is absorption pit method.

4.2. Harvesting Methods

Water falling on a flat rooftop should be made to run through a pipe connected to the roof and prevented from running off to the drainage on the roadside. This water should be filtered and stored in tanks constructed for the purpose of storing rainwater. In case of houses with sloped rooftops (R.C.C., Tiles and Asbestos), a gutter should be attached to the roof and the water collected in it should be made to go through a pipe attached to the end of the gutter. The water, which runs down from the pipe, should be filtered and stored in the rainwater harvest tanks. A good quality filter is needed for purifying the water. This is because, although the rainwater is basically pure, the water flowing from the rooftop usually will contain waste material in it.

4.2.1. For Individual Houses

Roof top rainwater can be diverted to the existing open / bore well. Along with this, rainwater available in the open spaces around the building may be recharged into the ground through the following simple effective methods.

- Percolation pits (Small houses)
- Recharge Trench (Big houses/Apartments)
- Recharge wells (Large buildings/industries)

Based on the size/area of the building and the underlying lithological nature of the formation the said methods may be used either individually or in combinations.To enhance nature recharge of rainwater avoid pavements since unpaved surfaces have more percolation rate.

In domestic Rooftop Rainwater Harvesting Systems rainwater from the house roof is collected in a storage vessel or tank for use during the periods of scarcity. Usually these systems are designed to support the drinking and cooking needs of the family at the doorstep. Such a system usually comprises a roof, a storage tank and guttering to transport the water from the roof to the storage tank. In addition, a first flush system to divert the dirty water which contains roof debris collected on the roof during non-rainy periods and a filter unit to remove debris and contaminants before water enters the storage tank are also provided.

The roof of the house is used as the catchment for collecting the rainwater. The style, construction and material of the roof effect its suitability as a catchment. Roofs made of corrugated iron sheet, asbestos sheet, tiles or concrete can be utilized as such for harvesting the rainwater. But thatched roofs are not suitable as it gives some colour to water and also the water carries pieces of roof material (such as palm leaves).

Gutters are channels fixed to the edges of roof all around to collect and transport the rainwater from the roof to the storage tank. Gutters can be prepared in semi-circular and rectangular shapes. Locally available material such as plain galvanised iron sheet can be easily folded to required shapes to prepare semi-circular and rectangular gutters. Semi-circular gutters of PVC material can be readily prepared by cutting the PVC pipes into two equal semi-circular channels. Bamboo poles can also be used for making gutters if they are locally available in sufficient quantity. Use of such locally available materials reduce the over all cost of the system.

Down pipe carries the rainwater from the gutters to the storage tank. Down pipe is joined with the gutters at one end, and the other end is connected to the filter unit of the storage tank. PVC or GI pipes of diameter 50 mm to 75 mm (2 inch to 3 inch) are commonly used for down-pipe. Bamboo can also be used wherever available in suitable size.

Debris, dirt and dust collect on the roofs during non-rainy periods. When the first rains arrive, this unwanted material will be washed into the storage tank. This caused contamination of water collected in the storage tank thereby rendering it unfit for drinking and cooking purposes. Therefore, a first flush system is incorporated in the Rooftop Rainwater Harvesting Systems to dispose off the 'first flush' water so that it does not enter the tank. There are two such simple systems. One is based on a simple manually operated arrangement, where by, the down pipe is moved away from the tank inlet and replaced again once the first flush water has been disposed. After the first rain is washed out through first flush pipe, the valve is closed to allow the water to enter the down pipe and reach the storage tank.

The filter unit is a container or chamber filled with filter media such as coarse sand, charcoal, coconut fiber, pebbles and gravels to remove the debris and dirt from water that enters the tank. The container is provided with a perforated bottom to allow the passage of water. The filter unit is placed over the storage tank. Commonly used filters are of two types. One is a ferrocement filter unit, which is comparatively heavy and the other is made of either aluminium or plastic bucket. The latter is readily available in market and has the advantage of ease in removing, cleaning and replacing.Another simple way of filtering the debris and dust particles that came from the roof along with rainwater is to use a fine cloth as filter media. The cloth, in 2 or 3 layers, can be tied to the top of a bucket or vessel with perforations at the bottom.

Storage tank is used to store the water that is collected form the Rooftops. Common vessels used for small scale water storage are plastic bowls, buckets, jerry cans, clay or ceramic jars, cement jars, old oil drums etc. For storing larger quantities of water the system will usually require a bigger tank with sufficient strength and durability.

The storage tank is provided with a cover on the top to avoid the contamination of water from external sources. The cover will be in dome shape having a raise of about 20-30 cm. in the middle. The dome is provided with two circular openings, one for manhole and another for accommodating the filter. A lid covers the manhole avoiding exposure of stored water to the outside environment. The storage tank is provided with pipe fixtures at appropriate places to draw the water, to clean the tank and to dispose of the excess water. They are named tap or outlet, drainpipe and over flow pipe respectively. PVC or GI pipes of diameter 20 mm to 25 mm are generally used for this purpose.

A small pit is dug in the ground, beneath the tap of the storage tank and constructed in brick masonry to make a chamber, so that a vessel could be conveniently placed beneath the tap for collecting water from the storage tank. A small hole is left at the bottom of the chamber, to allow the excess water to drain-out without stagnation. Size of collection pit shall be 60 cm x 60 cm x 60 cm.

4.2.1.1. *Percolation Pit Method*

4.2.1.1.1. *Recharge trench method*

- Similar to recharge pit but logitudinal in shape
- Size : 0.5 - 1 m wide 1 - 1.5 m depth. Length may vary from 1.0 - 5.0 km
- Filled with broken bricks /pebbles
- Suitable for sandy sub-soil area

4.2.1.1.2. *Recharge trench with bore method*

- Along the recharge trench boreholes has to be drilled
- Borehole interval 10 -15 feet
- Suitable for clay area
- Above structures are meant for area with large catchment like apartments/big complexes. RCC slab cover is optional
- Top portion may be filled with sand

4.2.1.1.3. *Recharge well method (shallow /small)*

- Meant for large areas with heavy run-off
- Size : 3' dia with 5' - 10' depth
- Constructed with brick / concrete rings
- Side walls must be perforated
- Bottom (1") is filled with broken brick
- Covered with RCC slab /manhole
- Suitable for sandy sub-soil area

- For areas with clay/hard rock formations recharge wells with large diameter (5 feet)/depth (10-15 feet) has to be constructed.(cost Rs.7,500/-)
- Run-off water diverted through a filter media preferably.
- One well is sufficient for areas with one ground extent.
- Rainwater from the open spaces may be diverted through small trenches.

4.2.1.2. *Bore Well with Settlement Tank*

- Roof top rainwater may also be diverted to a borewell
- Settlement / filter tank of required size has to be provided.
- Overflow water may be diverted to a percolation pit nearby.
- The rate of recharge through borewell is less effective than open wells.
- Defunct borewells may also be used.

4.2.1.3. *Openwell Method*

- Rainwater from the terrace is diverted to the existing open well using PVC pipes through a filter chamber.
- The minimum size of the filter chamber is 2" x 2" x 2" filled with broken bricks in the bottom and sand on the top. The chamber may be covered with RCC slab.

4.2.1.4. *Percolation Pit with Bore Method*

- A borehole to be drilled at the bottom of the percolation pit.
- Bore hole size : 150 - 300 mm dia with 10 -15 ft depth (approx.)
- Filled with broken bricks. Suitable for clay area.
- Above structures are meant for area with small catchment like individual houses.

- RCC slab cover is optional.
- Top portion may be filled with sand

4.2.2. Rainwater Harvesting in Group Houses

Utilise the open well if any, within the complex to divert the rainwater from the terrace into it. If not, construct a well for this purpose. The rainwater falling on the open space around the complex can be collected near the gate by providing a gutter with perforated lid. The collected water can be led through necessary piping arrangements into a recharge well of 1 metre dia and 5 metre deep.

4.3. Groundwater Dams

Groundwater dams are structures that intercept or obstruct the natural flow of groundwater and provide storage for water underground. They have been used in several parts of the world, notably India, Africa and Brazil. Their use is in areas where flows of groundwater vary considerably during the course of the year, from very high flows following rain to negligible flows during the dry season.

The basic principle of the groundwater dam is that instead of storing the water in surface reservoirs, water is stored underground. The main advantages of water storage in groundwater dams is that evaporation losses are much less for water stored underground. Further, risk of contamination of the stored water from the surface is reduced because as parasites cannot breed in underground water. The problem of submergence of land which is normally associated with surface dams is not present with sub-surface dams.

There are two main types of groundwater dam: the sub-surface dam and the sand storage dam.

4.3.1. Sub-surface Dam

A sub-surface dam intercepts or obstructs the flow of an aquifer and reduces the variation of the level of the groundwater table upstream of the dam. It is built entirely under the ground (figure 4.4).

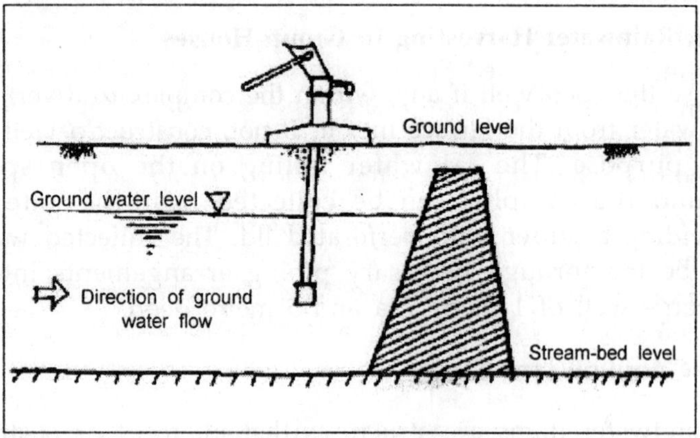

4.4. Sub-surface dam

4.3.2. Sand Storage Dam

The sand storage dam is constructed above ground. Sand and soil particles transported during periods of high flow are allowed to deposit behind the dam, and water is stored in these soil deposits (figure 4.5). The sand storage dam is constructed in layers to allow sand to be deposited and finer material be washed downstream (figure 4.6).

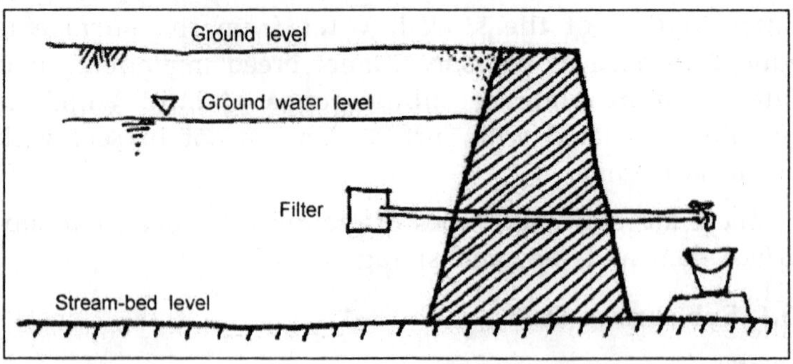

4.5. Sand storage dam

A groundwater dam can also be a combination of these two types. When constructing a sub-surface dam in a river bed, one can increase the storage volume by letting the dam wall rise over the surface, thus causing additional accumulation of sediments. Similarly, when a sand-storage dam is constructed it is necessary to excavate a trench in the sand bed in order to reach bedrock, which can be used to create a sub-surface dam too.

Groundwater dams are built across streams or valleys. A trench is dug across the valley or stream, reaching to the bedrock or other stable layer like clay. An impervious wall is constructed in the trench, which is then refilled with the excavated material.

Various materials may be used for the construction of groundwater dams. Materials should be waterproof, and the dam should be strong enough to withstand the imposed soil and water loads. Dams may vary from 2 to 10 metres high. Materials include compacted clay, concrete, stones and clay, masonry wall or plastic sheets.

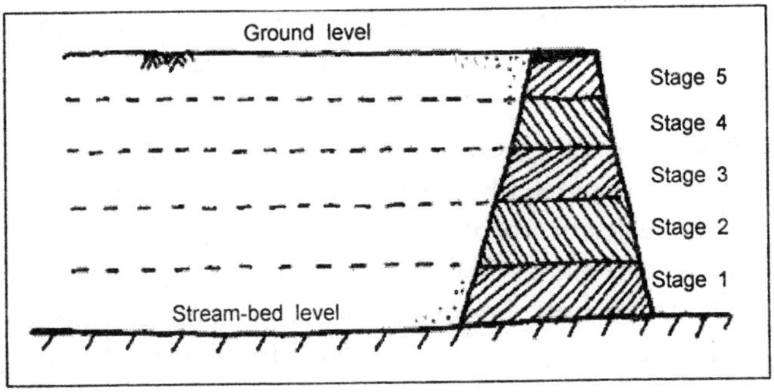

4.6. Stages of construction of a sand-storage dam

The reservoir is recharged during the monsoon period and the stored water can be used during the dry season. Excess water flows over the top of the dam to replenish aquifers downstream. Water may be obtained from the underground reservoir either from a well upstream of the dam or from a pipe, passing through the dam, and leading to a collection point downstream. Groundwater dams cannot be a universally applicable as these require specific conditions for functioning. The best sites for construction of groundwater dams are where the soil consists of sands and gravel, with rock or a permeable layer at a depth of a few metres. Ideally the dam should be built where rainwater from a large catchment area flows through a narrow passage.

4.4. Ferrocement Technology

One of the primary requirements of a water harvesting system is that of containers to store the harvested water in a hygienic condition. This need is more pronounced in high-rainfall areas, where it is more feasible to store water in containers for direct use, rather than for recharging the groundwater. Generally, in small domestic systems, the cost of constructing tanks with conventional materials like masonry or RCC is far more than that of the rainwater collection and piping component. Ferrocement can provide a low-cost and easy-to-build solution to the need for low-cost containers. This technology is particularly relevant for regions like Meghalaya, Arunachal Pradesh and Kerala, which have high frequency of rainfall.

Ferrocement consists of a thin sheet of cement mortar which is reinforced with a cage made of wire mesh and steel bars. Because ferrocement is structurally more effectient than masonry, the thickness of the walls of the container are as low as 10 to 15 mm. Ferrocement components can be casted in any shape using suitable mouolds. The technology is extremely simple to implement, and even semi-skilled workpersons can learn it with ease. Ferrocement requires only a few easily available materials—cement, sand, galvanised iron (GI) wire mesh, and mild steel (MS) bars—in small amounts compared to masonry and RCC.

The Structural Engineering Research Centre (SERC), Ghaziabad, has done research and development on a large number of low-cost structures and implements like water/

grain storage containers, irrigation channels, biogas digesters and septic tanks, primarily using ferrocement. SERC scientists are imparting training in ferrocement technology to rural artisans under the National Drinking Water Mission.

Ferrocement containers can be used to store grain and seed, apart from water. Tanks of 1000-2000 litre capacity can be constructed with ease, which are much cheaper than masonry, RCC or plastic tanks. These are easy to repair, and can be easily transported because of their sturdy nature. Such containers have been used on a wide scale since about the past 25 years in Thailand, Malaysia and some African countries. Ferrocement containers with capacity as much as 5000 litres have been constructed in Thailand.

The process of building a ferrocement container is very simple which users can do themselves, with some training. For a typical circular pot-shaped container, the only materials required are hessian cloth, chaff (waste from agricultural produce), GI wire mesh, MS bars, cement and sand.

To prepare the mould, the hessian cloth, stitched into a sack resembling the shape of the container, is filled with chaff that is compacted in layers. Dry leaves or dry grass can be used in place of chaff. Once the sack is filled with the filler material, it is beaten into the required shape by a wooden bat. A GI wire mesh is tied around the mould leaving sockets at suitable locations for inlet, overflow and cleaning pipes. Tying 6 mm diameter mild steel (MS) bars at wide intervals both horizontally and vertically strengthens the reinforcement cage.

Once the reinforcement cage is tied, cement mortar having cement-sand proportion of 1:2.5 or 1:3 (Table 4.1) is prepared, having water content equal to 0.45 times the volume of cement. The mortar is plastered in two layers along the wall thickness, the second layer being applied 24 hours after the first. The ferrocement wall normally has a thickness of 10 to 15 mm, depending on the volume of the container (Table 4.1).

Table 4.1 Parameters for various capacities of ferrocement containers

Capacity (ltrs)	Thickness of walls (mm)	Ratio cement:sand	Thickness of GI wire (guage)
400	10	1:3	26
600	10	1:3	24
900	12	1:2.5	24
1500	15	1:2.5	22

The cement mortar is applied ensuring a minimum clearance (cover) of 3 mm between the reinforcement mesh and the outer surfaces of the wall. The mould of the container is removed 24 hours after casting of the walls is completed, by removing the filler material, and the container can be brought into use after 10 days of wet curing.

References

Athawale, R. N. *Water Harvesting and Sustainable Supply in India,* published for Environment and Development Series, Centre for Environment Education, Rawat Publications, Jaipur and New Delhi. 2003.

Badiger, S.; Sakthivadivel, R.; Aloysius, N.; Sally, H. Preliminary Assessment of a Traditional Approach to Rainwater Harvesting and Artificial Recharging of Groundwater in Alwar District, Rajasthan, paper presented at the annual partners' meet of IWMI-Tata Water Policy Research Program, Anand, February 19-20. 2002.

Raju, K. C. B. Rainwater Harvesting for Groundwater Recharging: The Experience of SVRTI in Moti Rayan Village, Kachchh in Moench, M. (Ed.) *Groundwater Management: Supply Dominated Focus of Government and NGO Efforts,* Monograph, VIKSAT, Nehru Foundation for Development, Ahmedabad. 1995.

Sharma, P, *Ferrocement Bhandaran Patra (Ferrocement Storage Tanks)—National Drinking Water Mission,* SERC, Ghaziabad, 1990.

5 Rooftop Rainwater Harvesting

Rooftop rainwater harvesting is the technique through which rainwater is captured from the roof catchments and stored in reservoirs. Harvested rainwater can be stored in sub-surface ground water reservoir by adopting artificial recharge techniques to meet the household needs through storage in tanks.

The main objective of rooftop rainwater harvesting is to make water available for future use. Capturing and storing rainwater for use is particularly important in dryland, hilly, urban and coastal areas. In alluvial areas energy saving for 1m. rise in ground water level is around 0.40 kilo watt per hour.

Rooftop rainwater harvesting aimed:

1. To meet the ever increasing demand for water
2. To reduce the runoff which chokes storm drains
3. To avoid flooding of roads
4. To augment the ground water storage and control decline of water levels
5. To reduce ground water pollution
6. To improve the quality of ground water
7. To reduce the soil erosion
8. To supplement domestic water requirement during summer, drought etc.

5.1 Development of Rooftop Rainwater Harvesting

The application of an appropriate rainwater harvesting technology can make possible the utilisation of rainwater as a valuable and, in many cases, necessary water resource. Rainwater harvesting has been practised for more than 4, 000 years, and, in most developing countries, is becoming essential owing to the temporal and spatial variability of rainfall. Rainwater harvesting is necessary in areas having significant rainfall but lacking any kind of conventional, centralised government supply system, and also in areas where good quality fresh surface water or groundwater is lacking.

Annual rainfall ranging from less than 500 to more than 1 500 mm can be found in most Latin American countries and the Caribbean. Very frequently most of the rain falls during a few months of the year, with little or no precipitation during the remaining months. There are countries in which the annual and regional distribution of rainfall also differ significantly.

For more than three centuries, rooftop catchments and cistern storage have been the basis of domestic water supply on many small islands in the Caribbean. During World War II, several airfields were also turned into catchments. Although the use of rooftop catchment systems has declined in some countries, it is estimated that more than 500 000 people in the Caribbean islands depend at least in part on such supplies. Further, large areas of some countries in Central and South America, such as Honduras, Brazil, and Paraguay, use rainwater harvesting as an important source of water supply for domestic purposes, especially in rural areas.

5.2. Technical Aspects

A rainwater harvesting system consists of three basic elements: a collection area, a conveyance system, and storage facilities. The collection area in most cases is the roof of a house or a building. The effective roof area and the material used in constructing the roof influence the efficiency of collection and the water quality.

A conveyance system usually consists of gutters or pipes that deliver rainwater falling on the rooftop to cisterns or other storage vessels. Both drainpipes and roof surfaces should be constructed of chemically inert materials such as wood, plastic, aluminium, or fiberglass, in order to avoid adverse effects on water quality.

The water ultimately is stored in a storage tank or cistern, which should also be constructed of an inert material. Reinforced concrete, fiberglass, or stainless steel are suitable materials. Storage tanks may be constructed as part of the building, or may be built as a separate unit located some distance away from the building.

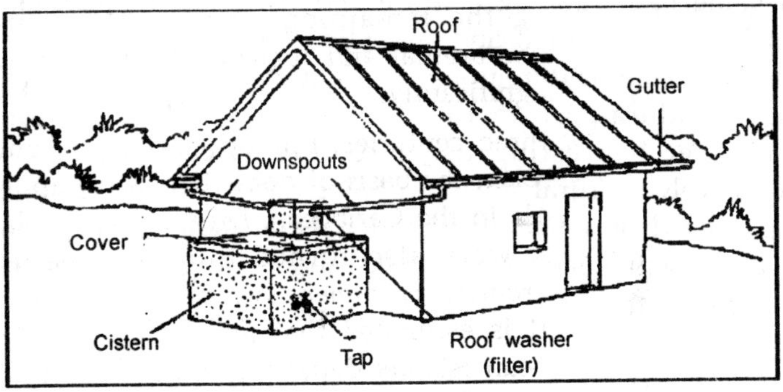

Figure 5.1. Schematic of a Typical Rainwater Catchment System.

All rainwater tank designs should include as a minimum requirement:

- A solid secure cover
- A coarse inlet filter
- An overflow pipe
- A manhole, sump, and drain to facilitate cleaning
- An extraction system that does not contaminate the water; e.g., a tap or pump
- A soakaway to prevent spilled water from forming puddles near the tank

Additional features are:

- A device to indicate the amount of water in the tank
- A sediment trap, tipping bucket, or other "foul flush" mechanism
- A lock on the tap
- A second sub-surface tank to provide water for livestock, etc.

5.3. Uses

Rainwater harvesting is used extensively in Latin America and the Caribbean, mainly for domestic water supply and, in some cases, for agriculture and livestock supplies on a small scale. In Brazil and Argentina, rainwater harvesting is used in semi-arid regions. In Central American countries like Honduras, Costa Rica, Guatemala, and El Salvador, rainwater harvesting using rooftop catchments is used extensively in rural areas. In Saint Lucia, storage tanks are constructed of a variety of materials, including steel drums, large polyethylene plastic tanks and underground concrete cisterns.

The Turks and Caicos Islands have a number of government-built, public rainfall catchment systems. Government regulations make it mandatory that all developers construct a water cistern large enough to store 400 l/m^2 of roof area.

Rooftop and artificially constructed catchments, such as the one at the former United States naval base on Eleuthera, are commonplace in the Bahamas. One settlement (Whale Cay) has a piped distribution system based on water captured from rooftops. On New Providence, most of the older houses collect rainwater from rooftops and store it in cisterns with average capacities of 7000l. Industries also use rooftop rainwater, and a preliminary assessment has been made of using Nassau International Airport as a catchment. In multistoried apartment buildings and other areas serving large concentrations of people (such as hotels and restaurants), water supplies are supplemented by water from rooftop catchment cisterns.

The Islas de la Bahía off the shores of Honduras meet a substantial portion of their potable water needs using rainwater from rooftop catchments. Similarly, rooftop catchments and cistern storage provide a significant water supply source for a small group of islands off the northern coast of Venezuela.

In a recent rural water-supply study, the continued use of rooftop and artificially constructed catchments was contemplated for those parts of rural Jamaica lacking access to river, spring, or well water sources. It is thought that more than 100 000 Jamaicans depend to a major extent on rainwater catchments.

5.4. Operation and Maintenance

Rainwater harvesting systems require few skills and little supervision to operate. Major concerns are the prevention of contamination of the tank during construction and while it is being replenished during a rainfall. Contamination of the water supply as a result of contact with certain materials can be avoided by the use of proper materials during construction of the system. For example, in Montserrat, where 95% of the houses in the medium to high density areas are roofed with oil-based bitumen shingles, consumers are strongly discouraged from using this source of supply for drinking purposes. Use of alternative roofing materials would have avoided this problem.

The main sources of external contamination are pollution from the air, bird and animal droppings, and insects. Bacterial contamination may be minimised by keeping roof surfaces and drains clean but cannot be completely eliminated. If the water is to be used for drinking purposes, filtration and chlorination or disinfection by other means (e.g., boiling) is necessary. The following maintenance guidelines should be considered in the operation of rainwater harvesting systems:

- A procedure for eliminating the "foul flush" after a long dry spell deserves particular attention. The first part of each rainfall should be diverted from the storage tank since this is most likely to contain undesirable materials which have accumulated on the roof and other surfaces between rainfalls. Generally, water captured during the first 10 minutes of rainfall during an event of average intensity is unfit for drinking purposes. The quantity of water lost by diverting this runoff is usually about 14l/m^2 of catchment area.
- The storage tank should be checked and cleaned periodically. All tanks need cleaning; their designs should allow for this. Cleaning procedures consist of thorough scrubbing of the inner walls and floors. Use of a chlorine solution is recommended for cleaning, followed by thorough rinsing.
- Care should be taken to keep rainfall collection surfaces covered, to reduce the likelihood of frogs, lizards, mosquitoes, and other pests using the cistern as a breeding ground. Residents may prefer to take care to prevent such problems rather than have to take corrective actions, such as treating or removing water, at a later time.
- Chlorination of the cisterns or storage tanks is necessary if the water is to be used for drinking and domestic uses. The Montserrat Island Water Authority constructed a non-conventional chlorination device with a rubber tube, plywood, a 1.2 m piece of PVC tubing, and a hose clip to chlorinate the water using chlorine tablets.
- Gutters and downpipes need to be periodically inspected and cleaned carefully. Periodic maintenance must also be carried out on any pumps used to lift water to selected areas in the house or building. More often than not, maintenance is done only when equipment breaks down.

- Community systems require the creation of a community organisation to maintain them effectively. Similarly, households must establish a maintenance routine that will be carried out by family members.

In some cases the rainwater is treated with chlorine tablets. However, in most places it is used without treatment. In such cases, residents are advised to boil the water before drinking. Where cistern users do not treat their water, the quality of the water may be assured through the installation of commercially available in-line charcoal filters or other water treatment devices. Community catchments require additional protections, including:

- Fencing of the paved catchment to prevent the entry of animals, primarily livestock such as goats, cows, donkeys, and pigs, that can affect water quality.
- Cleaning the paved catchment of leaves and other vegetative matter.
- Repairing large cracks in the paved catchment as a result of soil movement, earthquakes, or exposure to the elements.
- Maintaining water quality at a level where health risks are minimised. In many systems, this involves chlorination of the supplies at frequent intervals.

Problems usually encountered in maintaining the system at an efficient level include the lack of availability of chemicals required for appropriate treatment and the lack of adequate funding.

5.5. Choice of Roofing Materials

Roofs and gutters are made of a variety of materials. For most practical purposes we can exclude discussion of 'organic' roofs such as grass, reed and palm because they yield such dirty run-off that they are rarely used for rainwater harvesting. The common materials of interest are ceramic, cementitious, rock and metallic (plastic roofs being neither cheap nor durable). Contamination of water might arise from the roofing material itself or from substances that have accumulated on a roof or in a gutter.

Metal roofs are normally of treated steel or less commonly of aluminium. Aluminium is very inert unless in contact with very acid water. However the effect on health of ingesting tiny amounts of aluminium are controversial; there has been some debate in Europe about a possible link between such ingestion and the development of Alzheimer's Disease that causes premature senility. Corrugated steel roofing employs mild steel protected by hot-dip or electrolytic galvanising or by painting, since stainless steel is too expensive to use. Galvanising entails zinc compounds: fortunately zinc has a low toxicity, so that roof run-off water does not exceed WHO-permitted zinc levels.

Roof paints including bitumen may entail some risk to health and/or may impart unpleasant taste to roofwater and should probably be avoided for rainwater harvesting. More seriously although no one can now afford lead sheeting on roofs, localised lead 'flashing' is still used at joints. One study in Malaysia reported lead levels of up to 3.5 times WHO limits in roof runoff but this is not a general finding and seems to have arisen from lead in dust deposition rather than the roofing material since it reduced rapidly with storm duration. Not surprisingly the safety of water harvested from 'asbestos' (= asbestos-reinforced cement mortar) roofs has been queried, but the consensus is that the danger of developing cancer from ingested asbestos is very slight.

The danger from inhaled asbestos dust is however sufficiently high that working with asbestos sheeting, for example sawing it, without special protection is now generally banned in industrial countries. The iron in a rusting roof will also enter the runoff, but in such small quantities that it does not prejudice either health or taste.

Metal roofs are comparatively smooth and are therefore less prone to contamination by dust, leaves, bird-droppings and other debris than rougher tile roofs. They may also get hot enough to sterilise themselves. However contamination may be substantial on all roof types and it has been common for many years to design 'first-flush diverters' into rainwater harvesting systems. During a dry spell debris builds up on roofs, so that the initial run-off during the first following rainfall event can be full of sediment and highly turbid.

Overhanging trees, especially coconut palms, make this sediment problem worse, as well as increasing the likelihood of bird and rodent droppings. A common strategy therefore is to divert to waste the first say 5 litres of runoff at the beginning of each rain event. This can be done automatically using proprietary devices, or where the seasons are well defined it can be done manually by temporarily displacing the pipe connecting gutter to storage tank. If this first flush is excluded, we have a water source with modest levels of turbidity and typically medium levels of bacterial contamination (e.g. <10 FC per 100ml). Modern 'no-maintenance' separators, or more traditional screens, cloth or sand filters will reduce turbidity and contamination further and any good tank design will reduce it further still.

Not all tanks are however designed, made and maintained well. One can commonly see tanks which offer access to insects, lizards and rodents and which permit enough light to enter that algae can grow. Such tanks take longer to lower the contamination level of the entering flow and may even permit new infection for example by pathogens carried on the feet of cockroaches. Water abstraction is occasionally by lowered bucket—with all the opportunities for contamination that offers—and not uncommonly by a tap set too low in the tank so that tank-bottom sediment may be drawn into the outflow. However tests in even poorly designed tanks commonly give levels of bacterial contamination (rarely over 5 FC per 100ml) that compare well with those in competing water sources in developing countries. The technique of filling a rainwater harvesting tank then sealing it for a month or more produces excellent water quality. It seems it is possible even without such steps to meet the highest international standards for bacteria and dissolved substances with well-made rainwater harvesting systems incorporating effective prefiltration and careful in-tank flow guidance. Cleaning tanks, say annually, should improve water quality, provided any remaining disturbed sediment is allowed to resettle for several days before the tank is used again. With the best pre-tank separators however, the rate of entry of organic material is so low that (provided no photosynthesis occurs) such material can be entirely removed by aerobic bacterial action and no cleaning is required.

Water tanks are close to houses. Moreover they usually contain water during some or all of any dry season, a time when alternative breeding grounds for mosquitoes dry up. For both these reasons it is important that they do not act as significant breeding sites. The design of tanks *and guttering* to exclude insect breeding requires a mixture of common sense and professional engineering or entomological knowledge. It is common sense to so align gutters, and keep them clear of blockages, that they do not hold stagnant pools after rainfall finishes. It is engineering expertise or long experience that generates good designs for self-clearing gutters or filters. It perhaps requires entomological expertise to identify tank shapes that lower the chance of successful larval development.

Mosquito eggs are sufficiently small that they could pass through most filters with entry water: such filters cannot be very fine if they are to be able to handle the sudden and large flows during intense tropical rainfall of up to 1 millimetre per minute. Mosquito control is therefore a matter of preventing the entry and exit of adult insects and interfering with larval growth.

The former may be difficult to achieve if tank maintenance is poor or if users place greater importance on maximising tank inflow than on maintaining mosquito defences. It is therefore attractive to have the 'defence-in-depth' of larval control. This may take the form of active control with fish predators, surface oil films and suchlike but a more rewarding general policy is to starve larvae.

Maintaining darkness in a tank prevents photosynthesis and the growth of algae. Preventing the entry of suspended materials reduces the general nutrient levels supporting any biological chain. Research is underway into these factors and it seems likely that fairly straightforward measures can render a tank unsuitable for dry-season breeding of anopheles, aedes and culecine mosquitoes. Moreover, broadly speaking, if mosquitoes can be controlled it should be relatively easy to control larger disease vectors like cockroaches.

If a child falls into a tank, even if that child can swim, there is a real danger of drowning. Many existing tanks have no covers or easily displaced covers and stories of children deliberately bathing in free-standing rainwater harvesting tanks are to be heard. Perhaps of most concern are underground tanks whose covers have been opened for inspection, maintenance or even for drawing water. It is not normal to fence underground tanks, to extend them above the ground high enough to deter access by crawling babies or to socially control children from playing on them. However fencing and/or partial raising could have advantages including reducing danger of contamination by surface water and lowering the chance of cover damage by vehicles as well as reducing the risk of children or night-moving adults falling in. Accidents like drowning are most likely where a new technology is being introduced and therefore should be the particular concern of technology-change agents.

Every technology has its obscure and rarely-met failure and danger modes. Clearly any aspect of rainwater harvesting that involves human activity on high roofs, handling rusty metal or working 'underground' involves some risk of accident. A particular danger, known to have asphyxiated at least one builder recently, is the possibility of deoxygenation within a closed tank during the process of mortar setting and curing.

Finally, under health considerations, one might mention floods. If flood levels are higher than the entry point of rainwater harvesting tank entrances, there is the real danger of serious contamination of the stored water. This danger may be avoided by suitable tank location or by the permanent presence of a say slow-sand entry filter.

5.6. Participation of Government

The level of governmental participation varies in the countries of Latin America and the Caribbean. In some Caribbean islands, governments regulate the design of rainwater harvesting systems. In the U.S. Virgin Islands, the law requires that provision be made in the construction of all new buildings for the capture and storage of rainfall coming into contact with their roofs. The law requires that roofs be guttered and that cisterns be constructed having a volume that depends on the size of the roof, the intended use of the structure, and the number of floors. For a typical single-level, residential building, the law requires that 400 1 of storage be provided for each m^2 of roof area. Cistern construction is further regulated by the Virgin Islands Building Code to insure the structural integrity of these cisterns, which usually form an integral part of building foundations.

As of January 1, 1996, all new residences in Barbados are required to construct water storage facilities if the roof area or living area equals or exceeds 3 000 square feet. They will also be mandatory for all new commercial buildings with a roof area of 1 000 square feet or more. A rebate of $0.50 per gallon of installed tank capacity, up to the equivalent of 25% of the total roof area, will be given as an incentive by the Barbados Water Authority.

Cisterns are likely to continue to be a principal source of water for residences in several Caribbean islands. Even if mandatory requirements are removed, their use will remain widespread, as they provide a water supply that residents consider to be safe, sufficient, and inexpensive.

5.7. Costs

The cost of this technology varies considerably depending on location, type of materials used, and degree of implementation. In Brazil, the cost of a 30m^3 cistern in rural areas of the Northeast is around $900 to $1 000, depending on the material used. In the U.S. Virgin Islands, costs as low as $2 to $5/1000 are reported. Construction costs for underground cisterns can vary tremendously, based on the size and the amount of excavation required. In Saint Lucia, the average cost of a 1,500 plastic tank is $125.

In the Chaco region of Paraguay, two different types of cisterns have been used for rainwater harvesting: cisterns or storage tanks called *aljibes*, and cutwater cisterns called *tajamares*. The capital cost of a 30 m^3 cistern *(aljibe)* in Paraguay has been reported to be $2,000, while the construction of a 6,000 m^3 *tajamar*, including windmill-driven pumps and distribution piping, has been estimated at $8,400.

5.8. Effectiveness

Rainfall harvesting technology has proved to be very effective throughout several Latin American countries and most of the Caribbean islands, where cisterns are the principal source of water for residences. Cisterns are capable of providing a sufficient supply for most domestic applications.

The use of rainwater is very effective in lessening the demand on the public water supply system in the British Virgin Islands. It also provides a convenient buffer in times of emergency or shortfall in the public water supply. Also, because of the hilly or mountainous nature of the terrain in the majority of the British Virgin Islands, combined with dispersed housing patterns, rainfall harvesting appears to be the most practical way of providing a water supply to some residents. In many countries it is very costly, and in some cases not economically feasible, to extend the public water supply to all areas, where houses are isolated from one another or in mountainous areas.

Steep galvanised iron roofs have been found to be relatively efficient rainwater collectors, while flat concrete roofs, though highly valued as protection from hurricanes, are very inefficient. Rooftop catchment efficiencies range from 70% to 90%. It has been estimated that 1 cm of rain on 100 m^2 of roof yields 10 0001. More commonly, rooftop catchment yield is estimated to be 75% of actual rainfall on the catchment area, after accounting for losses due to evaporation during periods when short, light showers are interspersed with periods of prolonged sunshine. Likewise, at the other extreme, the roof gutters and downpipes generally cannot cope with rainfalls of high intensity, and excess water runs off the roof to waste during these periods.

This technology is suitable for use in all areas as a means of augmenting the amount of water available. It is most useful in arid and semi-arid areas where other sources of water are scarce.

5.9. Advantages

The major advantages of rooftop rainwater harvesting are given below:

- Rainwater harvesting provides a source of water at the point where it is needed. It is owner operated and managed.
- It provides an essential reserve in times of emergency and/or breakdown of public water supply systems, particularly during natural disasters.
- The construction of a rooftop rainwater catchment system is simple, and local people can easily be trained to build one, minimising its cost.
- The technology is flexible. The systems can be built to meet almost any requirements. Poor households can start with a single small tank and add more when they can afford them.

- It can improve the engineering of building foundations when cisterns are built as part of the substructure of the buildings, as in the case of mandatory cisterns.
- The physical and chemical properties of rainwater may be superior to those of groundwater or surface waters that may have been subjected to pollution, sometimes from unknown sources.
- Running costs are low.
- Construction, operation, and maintenance are not labour-intensive.

5.10. Disadvantages

The major disadvantages of rooftop rainwater harvesting are given below:

- The success of rainfall harvesting depends upon the frequency and amount of rainfall; therefore, it is not a dependable water source in times of dry weather or prolonged drought.
- Low storage capacities will limit rainwater harvesting so that the system may not be able to provide water in a low rainfall period. Increased storage capacities add to construction and operating costs and may make the technology economically unfeasible, unless it is subsidised by government.
- Leakage from cisterns can cause the deterioration of load bearing slopes.
- Cisterns and storage tanks can be unsafe for small children if proper access protection is not provided.
- Possible contamination of water may result from animal wastes and vegetable matter.
- Where treatment of the water prior to potable use is infrequent, due to a lack of adequate resources or knowledge, health risks may result; further, cisterns can be a breeding ground for mosquitoes.

— Rainfall harvesting systems increase construction costs and may have an adverse effect on home ownership. Systems may add 30% to 40% to the cost of a building.

— Rainfall harvesting systems may reduce revenues to public utilities.

5.11. Acceptability

Projects which involved the local community from the outset in the planning, implementation, and maintenance have the best chance of enduring and expanding. Those projects which have been predominantly run by local people have had a much higher rate of success than those operated by people foreign to an area, and those to which the community has contributed ideas, funds, and labor have had a greater rate of success than those externally planned, funded, and built. Successful rainwater harvesting projects are generally associated with communities that consider water supply a priority.

The attitudes toward the use of rainwater for domestic consumption differ. Some people, who depend on rainwater as their only source of supply, use it for all household purposes, from drinking and cooking to washing and other domestic uses. Other people, who have access to both rainwater and a public water supply, use rainwater selectively, for drinking or gardening or flushing toilets, and use the public water supply for other purposes. These varying attitudes are related to the level of education of the users as well as to their traditional preferences. Different sectors of the society need to be informed about the advantages of harvesting rainwater and the related safety aspects of its use, including the threat of mosquito problems and other public health concerns.

There is a need for the water quality aspects of rainwater harvesting to be better addressed. This might come about through:

— Development of first-flush bypass devices that are more effective and easier to maintain and operate than those currently available.

- Greater involvement of the public health department in the monitoring of water quality.
- Monitoring the quality of construction at the time of building. Other development needs include:
- Provision of assistance from governmental sources to ensure that the appropriate-sized cisterns are built.
- Promotion of rainwater harvesting as an alternative to both government- and private-sector-supplied water, with emphasis on the savings to be achieved on water bills.
- Provision of assistance to the public in sizing, locating, and selecting materials and constructing cisterns and storage tanks, and development of a standardised plumbing and monitoring code.
- Development of new materials to lower the cost of storage.
- Preparation of guidance materials (including sizing requirements) for inclusion of rainwater harvesting in a multi-sourced water resources management environment.

5.12. Success Stories

5.12.1. Rooftop Rainwater Harvesting in North Eastern States

North Eastern region, though endowed with high rainfall, suffers from water shortage during the dry months particularly in hilly areas and urban areas. Rainwater is considered to be a viable means to augment the water supply for domestic water needs. Considering this, roof top rainwater harvesting was taken in North-Eastern Region under Central Sector Scheme during IX Plan in Assam, Arunachal Pradesh, Meghalaya, Mizoram and Nagaland and under Fresh Water Year 2003 in Assam. Rainwater harvesting is a technique of collection and storage of rainwater in surface or in sub-surface aquifer before it is lost as surface run-off. The augmented resource is used in the time of need.

The methodology in harvesting rainwater from rooftop involves collection of rainwater from the roof through properly designed semi-circular gutters. The design and storage are decided based on the water to be harvested from the rainfall. The storage capacity has been decided on the daily requirement basis. The catchment area for the rooftop is taken as the plan area of the roof. Estimation of the water available from roof top rainwater harvesting is based on the catchment's area of the roof top considering. It requires a provision for flushing arrangement for the first showers to drain out the accumulated unwanted materials from roof top such dried and fallen leaves, bird's drop, etc.

From the roof top, water is collected through semi-circular gutters fitted along the edge of the sloping roof. Water from these semi-circular gutters are channelised through down flow pipes to storage tanks kept at the ground/underground. The gutters are fitted in such a way that the slope is maintained towards the storage tank. A bigger diameter of the semi-circular gutter is preferred as it will have a higher carrying capacity and will not overflow during heavy rain. Excess water may be routed to augment ground water resource. This may be achieved by diverting the excess water to abandoned dug wells or into a recharge pit. The size (or diameter) of the semi-circular gutters depends on the roof top/catchment area and the rain fall intensity. The number of structures constructed state-wise is depicted Table 5.1.

Table 5.1. State wise Rooftop Rainwater Harvesting Structures

State	*Central Sector Scheme*	*Scheme Fresh Water Year 2003*	*Total structures*	*Cost (in lakhs)*
Assam	12	1	13	23.22
Nagaland	42		42	116.43
Arunachal Pradesh	5		5	20.0
Meghalaya	6		6	20.32
Mizoram	35		35	28.0
Total	100		101	207.97

5.12.1.1. *Arunachal Pradesh*

Arunachal Pradesh is occupied mostly by hills (about 85% of the total geographical area) and 15% valley areas. Though it is heavily rain fed area, acute water scarcity is felt during lean period. To augment during lean period Roof Top Rainwater harvesting Scheme was selected the Ruksin Sub division of East Siang district. The details of schemes are given below.

Table 5.2. Rooftop Rainwater Harvesting Structures constructed in Arunachal Pradesh

Scheme	*Total fund allocated (Lakhs)*	*Implementing Agency*	*Status of work*	*Impact Assessment*
Roof Top Rainwater harvesting in Ruksin Sub division in East Siang district	20	IFCD Govt. of Arunachal Pradesh	5 structures completed	Water Tank capacity of each structure is 30,000 ltrs Total 170 persons are benefited through 5 schemes.

5.12.1.2. *Assam*

Under, Central Sector Scheme during IXth Plan, a total of 12 structures was constructed in Kamrup District. Hajo and Sonapur towns in Kamrup District have been selected for roof top rainwater harvesting schemes. 12 schools have been selected in two towns i.e. 6 in each town. The district is bestowed with abundant rainfall of over 2,000 mm spread over 7 months in a year.

Out of 12 structures, one structure i.e. Devalaya Prathamik Vidyalaya in Hajo was executed by CGWB, NER at the cost of Rs 84,252/-. However, other structures were executed by WAPCOS under overall supervision of CGWB, NER.

5.12.1.3. Harvesting under Fresh Water Year 2003

Under Fresh Water Year 2003, rooftop rainwater harvesting structure was constructed in Govt. building of Dept. of Geology and Mining in Dakshingaon, Kahilipara, Kamrup district, Assam. The structure with an underground storage capacity of 12,000 litres was constructed covering total roof area of 1134.69 sq. m at the total cost of Rs. 9.17 lakhs. 89,600 litres rainwater is harvested during monsoon period catering 390 persons for drinking, sanitation etc purposes.

5.12.1.3.1. Meghalaya

Meghalaya is predominantly a hilly terrain with limited valley areas at places. The average annual rainfall in the state is around 2,050 mm. Though, the state is endowed with the plenty of rainfall, it experiences shortage of drinking water during the summer. This may be due to high surface run off, as the area is hilly with steep slopes. The main water supply is dependent on spring, streams and waterfalls. To augment water supply, Central Ground Water Board has provided technical guidance for construction of six roof top rainwater harvesting structures in East Khasi Hills district. Out of these six, three schemes are in schools and three schemes in Govt. buildings.

5.12.1.3.2. Mizoram

The state constitutes a mountainous terrain with rugged topography represented by high longitudinal, parallel to sub parallel trending North South hills with moderate to steep slopes and narrow valleys. The hills are steep and separated by rivers creating deep gorges. The average annual rainfall in the states is 2200 mm. The major part of rainfall is lost as surface run-off. In spite of good rainfall in state there is acute shortage of water especially during the summer. Considering the physiography, rainfall, some suitable structures like check dams can be constructed for harvesting the surface run off. The present water resources of Mizoram can be augmented through development of springs and by constructing rainwater harvesting structures. The Central Ground Water Board has provided funds for construction of Rainwater harvesting structures of 35 selected sites in Aizawl district, Mizoram.

5.12.1.3.3. Nagaland

Nagaland is basically hilly terrain with small-elongated valleys constituting about only 6 percent of total geographical area. Acute scarcity of drinking water occurs during the lean period mainly in hilly areas, which comprise about 94 % of total area of the state. In order to augment water supply, construction of rainwater harvesting structures in 47 sites covering Kohima and Mokokchung districts have been done.

5.12.1.4. Bhoomual Samvardhan Puruskar

Bethany Society, Meghalaya was the only recipient of 'Bhoomijal Samvardhan Puraskar 2007 in North Eastern States. Ten Rooftop Rainwater Harvesting Tanks have been constructed in ten schools located in the East Khasi Hills, East, West and South Garo Hills district, Meghalaya. Twenty low cost toilets have been constructed, two in each school. There are separate toilets for girl students.

5.12.1.5. Benefits Accrued

- A total of 3,00,000 litres rainwater has been conserved in 10 tanks.
- 20 low cost toilets, 2 in each school.
- 836 students are availing of safe drinking water and sanitation facilities in the 10 schools.
- Village community came together to solve an urgent problem, Village community was trained to implement all aspects of a programme such as planning, purchase of materials, payment to workers, monitoring, social auditing, maintenance of records, share benefits according to norms set by the community.

- Awareness levels regarding water and sanitation were raised and community is demanding such services from their local representatives.
- Village community and school community were motivated to take charge of the programme and ensure its sustainability.

5.12.1.6. Type of Innovative Methodology Adopted

- Involvement of the community for the total implementation of the project.
- Gender sensitivity in formation of village level implementation committees.
- Transparency in all monetary transactions.
- Demonstration of both underground and above ground rainwater harvesting tanks.
- Use of rooftop rainwater harvesting technology.
- Use of two-pit low cost sanitation toilets.
- Use of social audit for monitoring.
- Collaboration among Ministry of Water Resources, CGWB, National Level NGO, Local NGO and local community to implement the programme.
- The scheme is under the direct charge of the Headmaster and Managing Committee of the respective school.

5.12.1.7. Benefit-Cost Ratio

- Approximately Rs. 80,000/- was spent on the construction of the tanks. Hence, the cost of per litre of water conserved has been worked out to be Rs. 2.66.
- The maintenance cost is almost negligible.

5.12.1.8. Sustainability

- The tanks have a life of 20 years and the only maintenance required is to ensure that the channels from the rooftops are well maintained and the tanks are cleaned and necessary chemicals are used for purification.

- The school and village communities were made the primary stakeholders.

5.12.1.9. *Awareness Generated*

- Awareness has been generated in all the ten villages and also some of the neighbouring villages on the value of rainwater harvesting, importance of portable drinking water and sanitation.
- Bethany Society has been approached by many villages and local community groups to expand the programme.

5.12.2.Rooftop Rainwater Harvesting in Schools of Karnataka

The Government of Karnataka through the Rural Development and Panchayath Raj department has launched a rooftop rainwater harvesting programme for 23,683 schools all across the state. The schools have been selected on the basis that they do not have any source of drinking water currently. The programme seeks to provide each student with 1.50 litres of drinking water through harvesting rainwater from rooftops of the schools. The construction component of the programme is being implemented by the Engineering Department of the various Zilla Panchayath or by the District Nirmithi Kendra.

Many schools presently do not have a reliable source of water for drinking and other use. The school rooftop rainwater harvesting system seeks to provide a source of water for all purpose such as toilet flushing, cooking, washing hands and feet before eating and after toilet use, hygiene and finally if the rainwater is treated well for drinking purpose. This is especially important in areas where there is Fluoride, Nitrate, iron or salt in the groundwater and therefore it is unfit for consumption. In these places the rainwater harvesting tank can provide mineral free water for consumption.

How much rainwater can be collected? This depends on the rooftop area, the size of the tank and the rainfall at that place. For example in a place where it rains 500 mm and the roof area is 100 square meters, the rainwater falling on the roof is 50,000 liters. Some amount of it will be absorbed by the roof and some amount will be lost in the collection process. If we assume 80% can be collected then 40,000 liters of rainwater is available for collection.

Depending on the size of the rainwater tank and the distribution of rainfall even a 3000 liter tank may be sufficient to collect all this 40,000 liters of rainwater. We must however keep on using the rainwater in the tank and not wait for summer to use it. By painting an information board, and keeping a small rain gauge in the school, a good school can involve students in monitoring rainfall, total rain in a year, water collected in the rainwater tank and teach them how to ensure good maintenance of the system.

5.12.2.1. Operation and Maintenance

The existing roof is made use of to collect rainwater. Since rainwater is pure as it falls from the sky it is necessary that the roof be kept clean for it to remain pure when it is collected. This means the roof will need to be swept and cleaned daily during the rainy season in the district.This should be carefully done by an adult equipped with the necessary implements such as a ladder, broom and a brush if necessary.Some schools will have shady trees to cover the roof. However leaves falling from the roof will cause blockage in the gutters and pipes. The leaves can also colour the water and cause it to decompose and smell. Therefore, roofs should invariably be completely cleaned of all leaves, dust, bird droppings etc. using a broom. Water should only be used if necessary as most times a dry sweeping with a broom will be enough. When cleaning the roof with water be careful to keep the first rain separator open so as not to allow the dirt to come into the filter and the tank.

The gutters of PVC collect the rainwater from the roof and transfer it to the filter. On sloping roofs, PVC gutters can pick up leaves, dust, small twigs and other organic matter. The gutters need to be cleaned regularly at least weekly once. During the rainy season the PVC gutters should be inspected and cleaned daily. The gutters are fixed to the roof or to the walls with clamps. The clamps hold the gutter or pipes to the wall or to the roof firmly and allow a small slope in the system to enable water to flow in one direction. The clamps sometimes may come off due to various reasons. The clamps should be fixed immediately whenever it is seen to be loose or when it has come off.At all times the PVC gutters or pipes should slope in the direction of the storage tank and not away from it. The PVC gutters should always have an end cap at the end where rainwater begins to flow in the direction of the tank.

PVC down pipes brings the water from the rainwater gutters or pipes vertically down. They should invariably be clamped firmly to the wall and should never be loosely fixed. Always check that the down pipe is firmly fixed and if necessary replace or tighten the clamp whenever necessary.

The first rain separator or a washout pipe as it is called, has a valve or an end cap to allow the first little amount of rainwater to be collected separately. This has most of the dust and dirt in it. The first rain separator also is used when the roof is being cleaned or when rainwater is NOT to be collected. It is important to ensure that the first rain separator is always kept in the closed position and never left open. After every rain it should be opened carefully and the waste water allowed to flow out. The pipe should then be cleaned and the valve or the end cap closed. Sometimes the first rain separator can get jammed due to the dirt or dust in it. In such a situation the valve or the end cap should be carefully replaced by a good plumber.

Where the roof of a school has lots of leaves falling from a tree or trees a conical leaf trap can be placed in the vertical down pipe. This has a mesh on top. The mesh prevents small leaves, twigs and other material from entering the pipe and blocking it or choking the filter. The material collected on the leaf trap if any must be removed at regular intervals and daily during the rainy season.

A gravel, sand and 'netlon' mesh filter is designed and placed on top of the storage tank. This filter is very important in keeping the rainwater in the storage tank clean. It removes silt, dust, leaves and other organic matter from entering the storage tank.The filter media should be cleaned daily after every rainfall event. Clogged filters prevent rainwater from easily entering the storage tank and the filter may overflow. The sand or gravel media should be taken out and washed before it is replaced in the filter.

The rainwater storage tank collects all the filtered rainwater and keeps it for future use. The storage tank is made above the ground and on a platform. It can also be an underground sump in some cases.The tank is invariably painted white on the outside. This is done to keep the water inside cool and prevent the growth of bacteria. Every year the tank must be white washed neatly. The tank also will be sealed from the top either with Cuddapah slabs or concrete slabs or any local stone. It must be ensured that the top cover is permanent and always fully covered. This will prevent the growth of algae or bacteria in the tank. In no case should it be opened. If there are small cracks in the joints they should be sealed with cement mortar immediately. Mosquitoes and dust should never be allowed in to the stored rainwater tank.

The tank should also be completely water tight. If there is any leak in the tank or even dampness, the problem should be addressed immediately with the help of a trained engineer.

The storage tank will have an overflow pipe from the top of the tank. In case of heavy rain, the overflow pipe will allow the excess rainwater to be safely disposed of without causing any flooding. The size of the overflow pipe should be the same as that of the inlet pipe. It will have a mesh at the bottom to prevent rats, squirrels and cockroach from coming in.The mesh should be checked weekly and if torn or open should be repaired or replaced immediately. It should also be ensured that the overflow water is drained away effectively to a pit, plant or storm water drain and not allowed to cause flooding.

A tap is provided in every tank to draw the rainwater out. Sometimes a tank can have more than one tap. Invariably it is found that children play with the pipe outlet or the tap and it is damaged. Children should be taught not to stand on the pipe or to play with the tap. A broken tap will result in the entire system going to waste as all the collected rainwater will flow out.If there is no tap on the tank or if it is broken, no rainwater will be collected in the tank when it rains. Ownership of the system should be created and the tap taken care of and inspected daily.If there is any leak in the tap, that too should be taken care of by replacing the washer or by getting a good plumber to repair it immediately.Where the taps are located the area will be cemented to drain out any waste water from the site. This drain out water will be lead into a pit or a plant whichever is available.

If the roof, the gutter, the first rain separator and the filter is kept clean, the collected rainwater will be crystal clear. This is an indication that good maintenance is being followed.If the water is however dirty in colour or it smells bad, then it means that the system is not being kept clean.Even if the water is clear and does not smell still it must be checked for micro-biological contamination. The checking should daily for the first one month and then weekly if the water is clear and not foul smelling. For this one must use a H_2S strip test bottle.

5.12.2.2. Water Quality Treatment

Though rainwater as it falls from the clouds is very pure, it does pick up dirt, dust and bacteria once it falls on the root. It is very necessary to therefore check the quality of the water before using it for consumption. Once it is established that the rainwater is not micro-biologically contaminated it can then be consumed directly. However if the H_2S strip test suggests that water has bacteria in it, it must then be treated before it can be used for drinking.

The method suggested for treating for bacteria is chlorination. Liquid chlorine or chlorine tablets are available for treatment of water. Depending on the volume of the rainwater in the tank, chlorine needs to be added to disinfect the water. Chlorination should be carried out every time there is rain and a fresh infusion of water into the tank. Using a chloroscope, residual chlorine of 0.20 mg/litre should be established before the water is used for drinking.

Another form of deactivating bacteria and making water fit for consumption is called SODIS – Solar disinfection of water. In this method, rainwater is kept in a PET bottle or a glass bottle in the sun for 6 hours. One side of the bottle is painted black. The black surface is kept on the ground. With a combination of UV disinfection and infra red heat sterilisation the water becomes fit for consumption. In cloudy weather the bottles need to be kept in the sun longer.

References

Hadwen, P. "Caribbean Islands: A Review of Roof and Purpose Built Catchments." In *Non-Conventional Water Resources Use in Developing Countries.* New York, United Nations, pp. 455-468. (Natural Resources/Water Series No. 22). 1987.

Proceedings of the First International Conference on Rainwater Cistern Systems. Honolulu, Hawaii, University of Hawaii, Water Resources Research Center. 1982.

Thomas, E.N. "The Artificial and Roof Rainwater Catches of Bermuda." Paper presented at the UN/CSC Seminar on Small Islands Water Problems. Barbados. 1980.

United Nations. Department of Technical Cooperation and Development. *Roof Catchments, Roof Coverings, Guttering and Downpipes.* New York. 1989.

UNEP. *Rain and Stormwater Harvesting in Rural Areas.* Nairobi, pp. 92-113. (Water Resources Series, 5) 1983.

6 In Situ Rainwater Harvesting

Rainwater harvesting is defined as a method for inducing, collecting, storing and conserving local surface runoff for agriculture in arid and semi-arid regions. Rainfall has four facets. Rainfall induces surface flow on the runoff area. At the lower end of the slope, runoff collects in the basin area, where a major portion infiltrates and is stored in the root zone. After infiltration has ceased, then follows the conservation of the stored soil water.

Various forms of rain water harvesting (RWH) have been used traditionally throughout the centuries. Some of the earliest agriculture, in the Middle East, was based on techniques such as diversion of "Wadi" flow (spate flow from normally dry water courses) onto agricultural fields. Other examples include the Negev desert, the desert areas of Arizona and Northwest Mexico and Southern Tunisia.The importance of traditional, small scale systems of rainwater harvesting in sub-Sahara Africa has recently been recognised. Simple stone lines are used, e.g. Burkina Faso and Mali; earth bunding systems in eastern Sudan, Kenya and the central rangelands of Somalia.

The potential of water harvesting for improved crop production received great attention in the 1970s and 1980s. This was due to the widespread droughts in Africa which left a trail of crop failures and a serious threat to human and livestock life. Consequently a number of water harvesting projects were set up in sub-Sahara Africa. The main objectives were to combat the effects of drought by improving plant production and in some areas rehabilitating abandoned and degraded land. However, few of the projects have succeeded in combining technical efficiency with low cost and acceptability to the local farmers or agro-pastoralists. This was partly due to the lack of technical "know how" but also often due to the selection of an inappropriate approach with regard to the prevailing socio-economic conditions.

In arid and semi-arid regions, where precipitation is low or infrequent during the dry season, it is necessary to store the maximum amount of rainwater during the wet season for use at a later time, especially for agricultural and domestic water supply. One of the methods frequently used in rainwater harvesting is the storage of rainwater *in situ.* Topographically low areas are ideal sites for *in situ* harvesting of rainfall. This technique has been used in the arid and semi-arid regions of northeastern Brazil, Argentina, and Paraguay, primarily for irrigation purposes. The *in situ* technology consists of making storage available in areas where the water is going to be utilised.

6.1. Technical Aspects

All rainfall harvesting systems have three components: a collection area, a conveyance system, and a storage area. In this application, collection and storage is provided within the landscape. Topographic depressions represent ideal collection and storage areas. In many situations, such areas are impermeable, being underlain by clay soils that minimise infiltration.

In Paraguay, areas of low topography used for rainwater storage are known as *tajamares. Tajamares* are constructed in areas with clay soils at least 3 m deep. The *tajamares* are served by distribution canals that convey water from the storage area to the areas of use. The collection and storage areas need to be fenced to avoid contamination by animals. This technology is usually combined with storage tanks built of clay.

Furrows may be used as an *in situ* means of storing harvested rainwater. They are built prior to or after planting to store water for future use by the plants. A variation on the use of topographic depressions to store rainfall, this method uses flattened trenches between the rows of crops to store water. Furrows may have mud dams or barriers every 2 m to 3 m along the row in order to retain water for longer periods of time and avoid excessive surface runoff and erosion. Raised beds may also be used to trap the water in the furrows, or uncultivated areas may be left between rows, spaced at 1 m apart, to assist in capturing rainwater falling on the land surface between furrows.

The *Guimares Duque* method was developed in Brazil during the 1950s, and uses furrows and raised planting beds, on which cross cuts to retain water are made using a reversible disk plow with at least three disks. The furrows are usually placed at the edge of the cultivation zone.

6.2. Usage

This technology has been extensively used in northeastern Brazil, in the Chaco region of Paraguay, and in Argentina. It can be used to augment the water supply for crops, livestock, and domestic use. With the mechanisation of agriculture, its use has diminished, but it is still recommended for regions where the volume of rainfall is small and variable. The approach used depends primarily on the availability of equipment, the nature of the agricultural and livestock practices, and the type of soil. Water stored in *tajamares* is normally used for livestock watering and may be used for domestic consumption after filtration and/or chlorination. Individual *tajamares* have also been used as a means of artificially recharging groundwater aquifers. *Tajamares* built in the Paraguayan Chaco have produced up to 6 800 m^3/yr for aquifer recharge.

6.3. Operation and Maintenance

This technology requires very little maintenance once the site is chosen and prepared. Maintenance is done primarily during the course of normal, day-to-day agricultural activities, and consists primarily of keeping the collection area free of debris and unwanted vegetation. Where only parts of the rows are cultivated, rotating the areas that are plowed will enable more efficient maintenance of the available storage area. This technology is simple and easy to use. Governmental organisations and the agricultural community generally work together to support and promote the *in situ* rainwater storage. Educational and information programmes should be provided to inform users of the benefits of this technology, and the means of implementing rainwater harvesting while preventing soil loss.

6.4. Costs

The costs of *in situ* rainwater collection systems are minimal. The main cost of this technology is in the equipment and labor required to build the fences and furrows. Table 6.1 shows representative costs reported for different methods of site preparation in cultivated areas of Brazil. Further, the construction cost of a *tajamar* in Paraguay has been reported at $4 500.

Table 6.1

Estimated Cost (S) of Different Site Preparation Methods for Rainwater Collection Areas in Agricultural Areas of Brazil

Method	*Basic Equipment*	*Animal Traction*	*Total*	*Hourly Cost of Implementation*
Flat terrain trenches	150.00	300.00	450.00	0.96
Post-planting furrows	80.00	300.00	380.00	0.90
Pre-planting furrows	180.00	70.00	250.00	0.90
Furrows with barriers	180.00	70.00	250.00	0.90
Inclined raised beds	1 500.00	1 000.00	2 500.00	12-15
Furrows in partial areas	100.00	80.00	180.00	0.70
Guimarães Duque method	–	–	–	12-15

6.5. Effectiveness

This technology increases water supply for irrigation purposes in arid and semi-arid regions. It promotes improved management practices in the cultivation of corn, cotton, sorghum, and many other crops. It also provides additional water supply for livestock watering and domestic consumption. This technology is applicable to low topographic areas in arid or semi-arid climates.

In situ rainwater technology has the following advantages:

- This technology requires minimal additional labor.
- It offers flexibility of implementation; furrows can be constructed before or after planting.
- Rainwater harvesting allows better utilisation of rainwater for irrigation purposes, particularly in the case of inclined raised beds.
- Rainwater harvesting is compatible with agricultural best management practices, including crop rotation.
- It provides additional flexibility in soil utilisation.
- Permeable *in situ* rainwater harvesting areas can be used as a method of artificially recharging groundwater aquifers.

The technology has the following disadvantages:

- *In situ* rainwater harvesting cannot be implemented where the slope of the land is greater than 5%.
- It is difficult to implement in rocky soils.
- Areas covered with stones and/or trees need to be cleared before implementation.
- The additional costs incurred in implementing this technology could be a factor for some farmers.
- It requires impermeable soils and low topographic relief in order to be effective.
- The effectiveness of the storage area can be limited by evaporation that tends to occur between rains.

6.6. Acceptability

In situ rainfall harvesting has been practised for many years by the agricultural communities of northeastern Brazil, Paraguay, and Argentina. Agricultural communities in other arid and semi-arid regions can readily improve their level of irrigation and increase their production yield using this technique. The equipment used in the construction of the furrows and storage areas must be improved. Relatively inexpensive plows and tractors can reduce the cost of implementation and contribute to the more widespread use of this technology by small farmers. New methods of soil conservation should be explored.

6.7. Use of in situ Tehchnology for Crop Production

6.7.1. Site and Technique Selection

Before selecting a specific technique, due consideration must be given to the social and cultural aspects prevailing in the area of concern as they are paramount and will affect the success or failure of the technique implemented. This is particularly important in the arid and semi-arid regions of Africa and may help to explain the failure of so many projects that did not take into account the people's priorities. In arid and semi-arid Africa, most of the population has experienced basic subsistence regimes which resulted over the centuries in setting priorities for survival. Until all higher priorities have been satisfied, no lower priority activities can be effectively undertaken.In addition to the socio-economic considerations, a water harvesting scheme will be sustainable if it also fulfils a number of basic technical criteria as shown in Figure 5.1. The chart shows the basic technical selection criteria for the different water harvesting techniques.

In crop production systems, RWH is composed of a runoff producing area normally called the catchment area (CA) and a runoff utilisation area normally called cropped basin (CB) The RWH systems for crop production are divided into different categories basically determined by the distance between CA and CB.

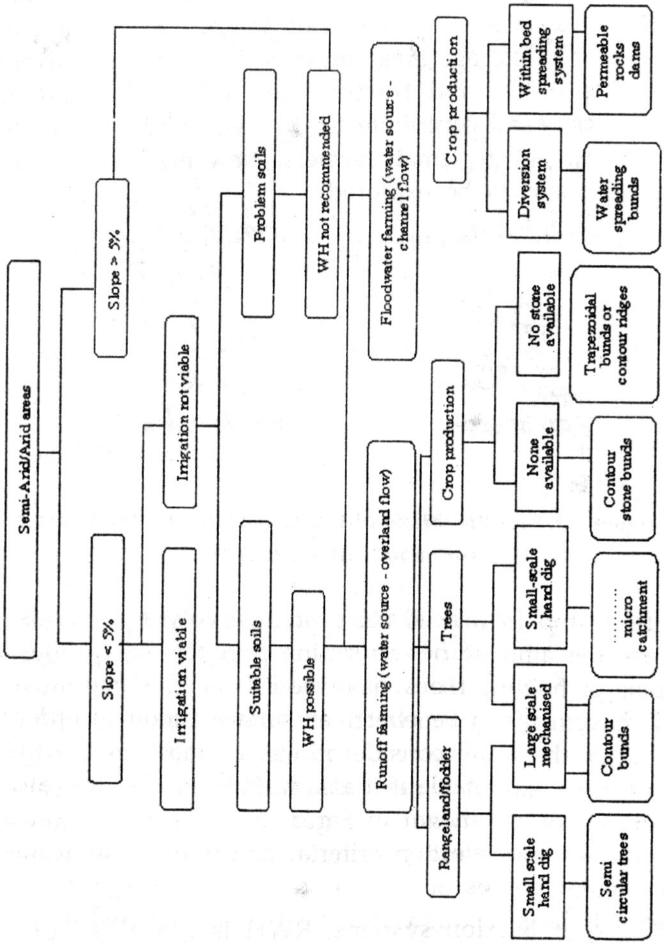

Figure 6.1. System selection

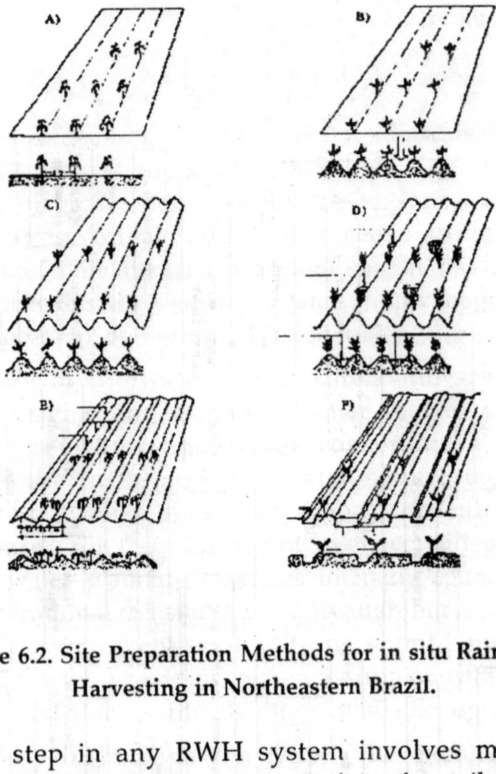

Figure 6.2. Site Preparation Methods for in situ Rainwater Harvesting in Northeastern Brazil.

The first step in any RWH system involves methods to increase the amount of water stored in the soil profile by trapping or holding the rain where it falls. This may involve small movements of rainwater as surface runoff in order to concentrate the water where it is wanted most. In-situ RWH is sometimes called water conservation and is basically a prevention of net runoff from a given cropped area by holding rain water and prolonging the time for infiltration. This system works better where the soil water holding capacity is large enough and the rainfall is equal or more than the crop water requirement, but moisture amount in the soil is restricted by the amount of infiltration and or deep percolation. The in-situ RWH is achieved mainly by the following means:

6.7.1.1. Deep Tillage

Tillage normally assists in increasing the soil moisture holding capacity through increased porosity, increasing the infiltration rates and reducing the surface runoff by providing surface micro-relief or roughness which helps in temporary storage of rain water, thus providing more time for infiltration.

Previous research results have shown that the depth of tillage is the most important factor controlling or affecting soil moisture characteristics. Deep tillage helps to increase porosity, reduce surface sealing of the soil and permits roots proliferation to exploit soil water and nutrients at deep horizons.

Significant reduction of surface runoff and increase in crop yields have been shown to occur with increased depth of tillage in Hombolo, Central Dodoma.

6.7.1.2. Contour Farming and Ridging

This is important where cultivation is done on slopes ranging from 3% and above. All farm husbandry practices such as tilling and weeding are done along the contours so as to form cross-slope barrier to the flow of water. Where this is not enough, it is complemented with ridges which are sometimes tied to create a high degree of surface roughness to enhance the infiltration of water into the soil.

Practices such as use of FYM, timely weeding and mulching are used to enhance water availability in the soil by improving the water holding capacity and reducing soil water evaporation.

6.7.2. Internal Catchment

This is a system where there is a distinct division of CA and CB but the areas are adjacent to each other. This system is mainly used for growing medium water demanding crops such as maize, sorghum, groundnuts and millet. The major characteristics of the system include:

– *Pitting*: These are small semi-circular pits dug to break the crusted soil surface. In West Africa where they are called 'Zay', the pits are about 30 cm in diameter and 20 cm deep. FYM is added in the pits thus permitting the concentration of water and nutrients. Seeds are planted in the middle of the pits. The same system is called Katumani pitting in Kenya. They are used in areas with rainfall of between 350-600 mm.

– *Strip catchment tillage*: This involves tilling strips of land along crop rows and leaving appropriate sections of the inter-row space uncultivated so as to release runoff. It is normally used where the slopes are gentle and the runoff from the uncultivated parts add water to the cropped strips.

– *Contour bunds*: This system consists of small trash, earth or stone embankments, constructed along the contour lines. The embankments trap the water flow behind the bunds allowing deeper infiltration into the soil. The height of the bund determines the net storage of the structure.The water is stored in the soil profile and above ground to the elevation of the bund or overflow structure. This is a versatile system for crop production in a variety of situations. They can be easily constructed but they are limited to availability of power (for earth moving), stones and trash.

– *Semi-circular bunds:* These are constructed in series in staggered formation. Runoff water is collected within the hoop from the area above it and impounded by the depth decided by the height of the bund and the position of the tips. Excess water is discharged around the tips and is intercepted by the second row and so on. Normally the semi-circles are of 4-12 m radius with height of 30 cm, base width of 80 cm, side slopes 1:1.5 and crest width of 20 cm. The percentage of enclosed area which is cultivated depends on the rainfall regime of the area. Basic requirements of the semi-circular bunds are:

  - ground slope must be less than 3%,
  - soil depth, at least 1 m,
  - average annual rainfall of at least 100 mm,
  - CBAR of at least 3:1 and
  - rainfall intensities of I60 equal to 50 mm/h for rainfall of P = 20%.

- *Meskat-type system*: In this system instead of having CA and CB alternating like the previous methods, here the field is divided into two distinct parts, the CA and CB, whereby the CB is immediately below the CA. In this system, the CA is treated either by removal of vegetation and compaction in order to increase the generation of runoff. The cropped basin (CB) is enclosed by a U-shaped bund to pond the harvested water. The CBAR is 2:1. It can be used for almost all cereal crops such as maize, sorghum and millet.
- *Land conservation aspects*: Micro-catchment approaches have a high potential for combining water harvesting with soil conservation. The main problem is that, in most projects there has been a bias towards promoting conservation rather than soil and water conservation with production. Conservation of both moisture and soil has two major advantages:
  - Due to increased crop yields, farmers will be more willing to implement and maintain the system;
  - The rapid vegetation development made possible by improved soil-moisture status, provides early protection to the soil against erosion.

  Micro-catchment rain-water harvesting, provides a good means for changing from soil conservation based on just runoff control to a focus on land husbandry integrating conservation and production.

6.7.3 External (Macro) Catchment

This is a system that involves the collection of runoff from large areas which are at an appreciable distance from where it is being used. This is sometimes used with intermediate storage of water outside the CB for later use as supplementary irrigation. It is difficult to differentiate this system from conventional irrigation systems but in this paper the system is called RWH as long as the water for harvesting is not available beyond the rainy season. This system involves harvesting of water from catchments of areas ranging from 0.1 ha to thousands of hectares either located near the cropped basin or long distances away. The catchment areas usually have slopes ranging from 5-50%, while the harvested water is used on cropped areas which are either terraced or on flat lands.When the catchment is large and located at a significant distance from the cropped area the runoff water is conveyed through structures of diversion and distribution networks. The most important systems include the following:

— *Hillside sheet/rill runoff utilisation*: In this system, runoff which occurs on hill-tops (with stone outcrops), sloping grounds, grazing lands or other compacted areas flow and naturally collect on low lying flat areas. In many areas farmers grow their crops on the wetted part of the landscape and use the runoff without any further manipulation or management.However, where the runoff is not high, bunds are constructed on the cropped area in order to form earth basins which assist in holding the water and increasing infiltration into the soil. These bunds are important when the cropped area is not at the bottom of the landscape. However earth basins are used to facilitate the distribution of the water even if the cultivated area is on flat land. Several designs of these earth basins are used and sometimes are mentioned as types of RWH systems by themselves. These include, for example trapezoidal basins bunded on three sides, rectangular basins bunded on three sides e.g. Teras, and cultivated basins bunded on all the four sides with only small inlets and overflow spillway, e.g. 'majaruba'.

- *Floodwater harvesting within the stream bed*: This is a system that uses barriers such as permeable stone dams to block the water flow and spread it on the adjacent plain and enhance infiltration. The wetted area is then used for crop production.
- *Ephemeral Stream Diversion*: This system involves means for diverting water from its natural ephemeral stream and conveying it to arable cropping areas. There are two main methods of diverting and distributing the water.In the first system, the cultivated field close to the ephemeral stream is first divided into open basins using either trapezoidal, semi-circular bunds or rectangular. By means of a weir, water is diverted from the stream into the top most basin. The water fills the basin and the surplus spills to the next basin and so on until the whole farm is fully wetted.In this system, one intake point can only be used by a single farm which must be relatively close to the source. In the second system, the field is divided into a closed rectangular basin such as "majaruba" and the water is diverted using a weir and a series of channels to deliver the water to the basins. The system works using the same principles of surface flood irrigation and it can therefore serve several farms which may be located far away from the intake.
- *RWH with storage*: Sometimes macro-catchment RWH, produces high volumes of runoff that can not be stored in the soil profile. In such circumstances, the harvested water is stored in dams or water holes. Small dams are normally constructed in rolling topography where creeks can be found and the dams are constructed across them.

  Water holes are storage ponds dug in a flat terrain and they are normally referred to in their Spanish name "Charco dams". In India they are called 'tanks'. They are normally used to store runoff generated from hillside catchments with sheet or rill flow. The system requires methods for controlling siltation especially if the area is prone to soil erosion, evaporation, and seepage losses especially if the subsoil is sandy.

— *Land conservation aspects*: Land conservation role of macro-catchment RWH is much difficult to visualise. In reality macro-catchment RWH system condemn one part of the land (the catchment) for the benefit of another (the cropped basin). This is occurring naturally on the catena, where fine particles, nutrients and organic matter are disproportionately lost in the eroded fraction from catchment areas. This is referred to as the "enrichment factor", and is the amount by which the eroded material is richer in nutrients than is the soil from which it is taken. For this reason careful planning of macro-catchment systems is necessary so as to avoid the degradation of land from which runoff is occurring.

References

Boers, Th M, and Ben-Asher, J. "A review of rainwater harvesting". *In Agric. Water Management* 5:145-158. 1982.

Evenari, M, Shana, L and Tadmor, N H. *The Neger, the challenges of a desert*. Havard University Press. Cambridge, Mass. 1971.

Zaunderer, J and Hutchinson, C F. A review of water harvesting techniques of the Arid Southwestern US and North Mexico. Working paper for the World Bank's Sub-Sahara Water Harvesting Study. 1988.

Pacey, A and Cullis, A. *Rainwater harvesting; The Collection of rainfall and runoff in rural areas*. IT Publication, London. 1986.

Hudson, N. Soil and Water Conservation in semi-arid areas. FAO Soils Bulletin No. 57. FAO, Rome. 172 pp. 1987.

SWMRG, Evaluation and Promotion of Rainwater harvesting in Semi-Arid areas of Tanzania Research Project. 2nd interim Technical report. Sokoine University of Agriculture, Morogoro. 80 pp. 1995.

7 Water Quality and Treatment

Rainwater harvesting is one of the world's oldest water supply methods and it is currently enjoying a revival in popularity. Rain has long been valued as a superior quality water because it is soft, free of sodium and chemicals, and is excellent for landscape use. Collecting rain also reduces flooding and can help utilities reduce peak demands during summer months.

Rainwater is relatively free from impurities except those picked up by rain from the atmosphere, but the quality of rainwater may deteriorate during harvesting, storage and household use. Wind-blown dirt, leaves, faecal droppings from birds and animals, insects and contaminated litter on the catchment areas can be sources of contamination of rainwater, leading to health risks from the consumption of contaminated water from storage tanks.

Poor hygiene in storing water in and abstracting water from tanks or at the point of use can also represent a health concern. However, risks from these hazards can be minimized by good design and practice. Well designed rainwater harvesting systems with clean catchments and storage tanks supported by good hygiene at point of use can offer drinking-water with very low health risk, whereas a poorly designed and managed system can pose high health risks.

Microbial contamination of collected rainwater indicated by E. coli (or, alternatively, thermotolerants coliforms) is quite common, particularly in samples collected shortly after rainfall. Pathogens such as Cryptosporidium, Giardia, Campylobacter, Vibrio, Salmonella, Shigella and

Pseudomonas have also been detected in rainwater. However, the occurrence of pathogens is generally lower in rainwater than in unprotected surface waters, and the presence of non-bacterial pathogens, in particular, can be minimized.

Higher microbial concentrations are generally found in the first flush of rainwater, and the level of contamination reduces as the rain continues. A significant reduction of microbial contamination can be found in rainy seasons when catchments are frequently washed with fresh rainwater. Storage tanks can present breeding sites for mosquitoes, including species that transmit dengue virus.

Rainwater is slightly acidic and very low in dissolved minerals; as such, it is relatively aggressive. Rainwater can dissolve heavy metals and other impurities from materials of the catchment and storage tank. In most cases, chemical concentrations in rainwater are within acceptable limits; however, elevated levels of zinc and lead have sometimes been reported. This could be from leaching from metallic roofs and storage tanks or from atmospheric pollution.

Rainwater lacks minerals, but some minerals, such as calcium, magnesium, iron and fluoride, in appropriate concentrations are considered very essential for health. Although most essential nutrients are derived from food, the lack of minerals, including calcium and magnesium, in rainwater may represent a concern for those on a mineral-deficient diet. In this circumstance, the implications of using rainwater as the primary source of drinking-water should be considered. The absence of minerals also means that rainwater has a particular taste or lack of taste that may not acceptable to people used to drinking other mineral-rich natural waters.

7.1. System Risk Assessment

Important factors in collecting and maintaining good quality rainwater include proper design and installation/ construction of rainwater harvesting systems. Materials used in the catchment and storage tank should be suitable for use in contact with drinking-water and should be non-toxic to humans.

Rainwater can be harvested using roof and other above-ground catchments and stored in tanks for use. The roof catchment is connected with a gutter and down-pipe system to deliver rainwater to the storage tank. The quality of rainwater is directly related to the cleanliness of catchments, gutters and storage tanks. Rooftop catchment surfaces collect dust, organic matter, leaves and bird and animal droppings, which can contaminate the stored water and cause sediment buildup in the tank. Care should also be taken to avoid materials or coatings that may cause adverse taste or odour, and some metals can dissolve to give high concentrations in water. Regular cleaning of catchment surfaces and gutters should be undertaken to minimize the accumulation of debris. Wire meshes or inlet filters should be placed over the top of down-pipes to prevent leaves and other debris from entering storages. These meshes and filters should be cleaned regularly to prevent clogging.

The first flush of rainwater carries most contaminants into storages. A system is, therefore, necessary to divert the contaminated first flow of rainwater from roof surfaces. Some devices and good practices are available to divert the first foul flush of rainwater. Automatic devices that prevent the first 20–25 litres of runoff from being collected in storages are recommended. If diverters are not available, a detachable down-pipe can be used manually to provide the same result. Even with these measures in place, storages will require periodic cleaning to remove sediment.

Storages without covers or with unprotected openings will encourage mosquito breeding, and sunlight reaching the water will promote algal growth. Covers should be fitted, and openings need to be protected by mosquito-proof mesh. Cracks in the tank and withdrawing of water using contaminated pots can contaminate stored water. Storages should preferably be fitted with a mechanism such as a tap or outlet pipe that enables hygienic abstraction of water. Some households incorporate cartridge filters or other treatments at the point of consumption to ensure better quality of drinking-water and reduce health risk.

It is worth noting that owners of rainwater harvesting systems who supply all domestic needs essentially become owners of their "water supply systems," responsible for routine maintenance, including filter and lamp replacement, leak repair, monitoring of water quality, and system upgrades.

The rainwater harvesting system owner is responsible for both water supply and water quality. Maintenance of a rainwater harvesting system is an ongoing periodic duty, to include:

- monitoring tank levels,
- cleaning gutters and first-flush devices,
- repairing leaks,
- repairing and maintaining the system, and
- adopting efficient water use practices.

In addition, owners of potable systems must adopt a regimen of:

- changing out filters regularly,
- maintaining disinfection equipment, such as cleaning and replacing ultraviolet lamps, and
- regularly testing water quality.

7.2. Factors Affecting Water Quality

The raindrop as it falls from the cloud is soft, and is among the cleanest of water sources. Use of captured rainwater offers several advantages. Rainwater is sodium-free, a benefit for persons on restricted sodium diets. Irrigation with captured rainwater promotes healthy plant growth. Also, being soft water, rainwater extends the life of appliances as it does not form scale or mineral deposits.The environment, the catchment surface, and the storage tanks affect the quality of harvested rainwater. With minimal treatment and adequate care of the system, however, rainfall can be used as potable water, as well as for irrigation.

The falling raindrop acquires slight acidity as it dissolves carbon dioxide and nitrogen. Contaminants captured by the rain from the catchment surface and storage tanks are of concern for those intending to use rainwater as their potable water source. The catchment area may have dust, dirt, fecal matter from birds and small animals, and plant debris such as leaves and twigs.

7.2.1. pH (acidity/alkalinity)

As a raindrop falls and comes in contact with the atmosphere, it dissolves naturally occurring carbon dioxide to form a weak acid. The resultant pH is about 5.7, whereas a pH of 7.0 is neutral. (A slight buffering using 1 tablespoon of baking soda to 100 gallons of water in the tank will neutralise the acid, if desired. Also, a concrete storage tank will impart a slight alkalinity to the water.)

7.2.2. Particulate Matter

Particulate matter refers to smoke, dust, and soot suspended in the air. Fine particulates can be emitted by industrial and residential combustion, vehicle exhaust, agricultural controlled burns, and sandstorms. As rainwater falls through the atmosphere, it can incorporate these contaminants.

7.2.3. Chemical Compounds

Information on chemical constituents can also be found on the TCEQ Air Quality website. In agricultural areas, rainwater could have a higher concentration of nitrates due to fertiliser residue in the atmosphere. Pesticide residues from crop dusting in agricultural areas may also be present. Hard water has a high mineral content, usually consisting of calcium and magnesium in the form of carbonates. In industrial areas, rainwater samples can have slightly higher values of suspended solids concentration and turbidity due to the greater amount of particulate matter in the air.

7.2.4. Catchment Surface

When rainwater comes in contact with a catchment surface, it can wash bacteria, molds, algae, fecal matter, other organic matter, and/or dust into storage tanks. The longer the span of continuous number of dry days (days without rainfall), the more catchment debris is washed off the roof by a rainfall event

7.2.5. Tanks

The more filtering of rainwater prior to the storage tanks, the less sedimentation and introduction of organic matter will occur within the tanks. Sedimentation reduces the capacity of tanks, and the breakdown of plant and animal matter may affect the colour and taste of water, in addition to providing nutrients for microorganisms.

Most storage tanks are equipped with manholes to allow access for cleaning. Sediment and sludge can be pumped out or siphoned out using hose with an inverted funnel at one end without draining the tank annually.

Multiple linked tanks allow one tank to be taken off line for cleaning by closing the valve on the linking pipe between tanks.

7.3. Water Treatment

The cleanliness of the roof in a rainwater harvesting system most directly affects the quality of the captured water. The cleaner the roof, the less strain is placed on the treatment equipment. It is advisable that overhanging branches be cut away both to avoid tree litter and to deny access to the roof by rodents and lizards.

For potable systems, a plain galvanised roof or a metal roof with epoxy or latex paint is recommended. Composite or asphalt shingles are not advisable, as toxic components can be leached out by rainwater.

To improve water quality, several treatment methods are discussed. It is the responsibility of the individual installer or homeowner to weigh the advantages and disadvantages of each method for appropriateness for the individual situation. A synopsis of treatment techniques is shown in Table 7.1.

7.3.1. Chlorination

Chlorination is mentioned here more for its historical value than for practical application. Chlorine has been used to disinfect public drinking water since 1908, and it is still used extensively by rainwater harvesters in Hawaii, the U.S. Virgin Islands, and in older rainwater harvesting systems in Kentucky and Ohio. Chlorine must be present in a concentration of 1 ppm to achieve disinfection. Liquid chlorine, in the form of laundry bleach, usually has 6 percent available sodium hypochlorite. For disinfection purposes, 2 fluid ounces (¼ cup) must be added per 1,000 gallons of rainwater. Household bleach products, however, are not labelled for use in water treatment by the Food and Drug Administration. A purer form of chlorine, which comes in solid form for swimming pool disinfection, is calcium hypochlorite, usually with 75 percent available chlorine. At that strength, 0.85 ounces by weight in 1,000 gallons of water would result in a level of 1 ppm. In either case, it is a good idea to carefully dilute the chlorine source in a bucket of water, and then stir with a clean paddle to hasten mixing. Chlorine contact times are show in Table 7.2.

The use of chlorine for disinfection presents a few drawbacks. Chlorine combines with decaying organic matter in water to form trihalomethanes. This disinfection by-product has been found to cause cancer in laboratory rats. Also, some users may find the taste and smell of chlorine objectionable. To address this concern, an activated carbon filter may be used to help remove chlorine.

Table 7.1. Treatment Techniques

Method	*Location*	*Result*
Treatment		
Screening		
Leaf screens and strainers	gutters and downspouts	prevent leaves and other debris from entering tank
Settling		
Sedimentation	within tank	settles out particulate matter
Activated charcoal	before tap	removes chlorine
Filtering		
Roof washer	before tank	eliminates suspended material
In-line/multi-cartridge	after pump	sieves sediment
Activated charcoal	after sediment filter	removes chlorine, improves taste
Slow sand	separate tank	traps particulate matter
Microbiological treatment /Disinfection		
Boiling/distilling	before use	kills microorganisms
Chemical treatments (Chlorine or Iodine)	within tank or at pump (liquid, tablet, or granular) before activated charcoal filter	kills microorganisms
Ultraviolet light	after activated charcoal filter, before tap	kills microorganisms
Ozonation	after activated charcoal filter, before tap	kills microorganisms
Nanofiltration	before use; polymer membrane (pores 10^{-3} to 10^{-6} inch)	removes molecules
Reverse osmosis	before use: polymer membrane (pores 10-9 inch)	removes ions (contaminants and microorganisms)

Table 7.2. Contact Time with Chlorine

Water pH	*Water temperature*		
	50 F or warmer	*45 F*	*40 F or colder*
	Contact time in minutes		
6.0	3	4	5
6.5	4	5	6
7.0	8	10	12
7.5	12	15	18
8.0	16	20	24

Chlorine does not kill Giardia or Cryptosporidium, which are cysts protected by their outer shells. Persons with weakened or compromised immune systems are particularly susceptible to these maladies. To filter out Giardia and Cryptosporidum cysts, an absolute 1-micron filter, certified by the NSF, is needed.

7.3.2. UV Light

Ultraviolet (UV) disinfection uses a UV light source, which is enclosed in a transparent protective sleeve. It is mounted so that water can pass through a flow chamber, and UV rays are admitted and absorbed into the stream. When ultraviolet energy is absorbed by the reproductive mechanisms of bacteria and viruses, the genetic material (DNA/RNA) is rearranged and they can no longer reproduce. They are therefore considered dead and the risk of disease has been eliminated.

UV-rays are energy-rich electromagnetic rays that are found in the natural spectrum of the sunlight. They are in the range of the invisible short wave light having a wavelength ranging from 100 to 400 nm (1 nanometre = 10-9m).

UV light has been used in for disinfection of water since the early 1900s, and its use has now become common practice now. utilities. Bacteria, virus, and cysts are killed by exposure to UV light. The water must go through sediment filtration before the ultraviolet light treatment because pathogens can be shadowed from the UV light by suspended particles in the water. In water with very high bacterial counts, some bacteria will be shielded by the bodies of other bacteria cells.

UV, like distillation, disinfects water without adding chemicals, and therefore possesses some of the same benefits as distillation. It does not create new chemical complexes, nor does it change the taste or odor of the water, and does not remove any beneficial minerals in the water.

Ultraviolet devices are most effective when the water has already been partially treated, and only the cleanest water passes through the UV flow chamber. Niagara+ UV Purifiers use both a sediment and a carbon filter to clean the water prior to passing it through the UV light, to provide complete water quality solutions. Ultraviolet light is a natural, cost effective, environmentally friendly disinfection process for use in homes where healthy water is a concern.

7.3.2.1. UV Sterilizer

A typical UV (ultraviolet) sterilizer consists of a UV-emitting lamp mounted within a quartz sleeve (a special type of glass tube that does not block UV) which is in turn mounted within a stainless-steel tube. Water flowing between the quartz sleeve and the stainless-steel tube is exposed to UV energy which theoretically kills any microorganisms in the water. Better units also monitor the amount of UV energy that reaches the stainless-steel tube.

The design and operation of our UV sterilizer is completely different: two UV lamps and a quartz sleeve are all mounted inside a dry enclosure, one lamp on either side of the quartz sleeve. Water flows through the inside the quartz sleeve and is exposed to UV energy emitted by both lamps. A sophisticated monitoring system measures the UV output from the lamps and compares it with the amount of UV energy that passes through the water. This approach exposes microorganisms in the water to significantly more UV energy, eliminates shadowing behind larger suspended particles, and accurately determines the exact UV exposure.

7.1. UV sterilizer

The problem with using typical UV sterilizers for rainwater is that although rainwater is a relatively high-quality water source, it can have elevated levels of dissolved organics, iron compounds, and phosphates that absorb UV energy. It is not uncommon to see the percent of UV energy transmission, called the UVT, drop below 70% to 75%, the minimum UVT required for ordinary UV sterilizers. Our high-intensity sterilizer is designed to disinfect when UVT levels are as low as 50%, a level rarely seen in rainwater.

Over time, the quartz sleeves in UV sterilizers become coated with a mineral deposit that inhibits UV transmission. The simplest UV sterilizers require periodic manual disassembly and cleaning, a difficult and messy process that occasionally results in a broken sleeeve. Better UV sterilizers utilize a sleeve wiping device, but these are not completely effective and still require a regular maintenance schedule. UV sterilizers with UV monitors are also prone to problems caused by mineral deposits on the sensors which are exposed to water. Our UV sterilizer features a motor-driven internal wiper that requires no user action, and since the UV monitoring system in our sterilizer is not in contact with water it does not require cleaning.

For general rainwater disinfection, either interior or exterior, we supply a version with a 1" ports suitable for flow rates up to 15 GPM and a version with 1.5" ports suitable for flow rates up to 30 GPM. Contact us for sizing for drinking water systems. Both units include a sophisticated microprocessor-controlled monitoring system, a 24v output to operate an optional solenoid valve to stop water flow in the event of insufficient sterilization, plus an automatic bleed system to prevent overheating that can reduce UV output.

7.3.3. Ozonation

Ozone has been used in Europe for water treatment since early in the 20th century. Initial applications were to disinfect relatively clean spring or well water, but they increasingly evolved to also oxidize contaminants common to surface waters. Since World War II, ozonation has become the primary method to assure clean water in Switzerland, West Germany and France. More recently, major fresh water and waste water treatment facilities using ozone water treatment methods have been constructed throughout the world.

In the field of creative ozone water treatment, BiOzone designs, manufactures, and installs the finest ozone generator systems produced today. Each component interfaces harmoniously with the others to achieve the most cost effective and optimum performance. These process trains are installed around the world in a variety of industries. The solutions we offer are nowhere else available in the world.

Ozone is an unstable gas comprising of three oxygen atoms, the gas will readily degrade back to oxygen, and during this transition a free oxygen atom, or free radical is form. The free oxygen radical is highly reactive and short lived, under normal conditions it will only survive for milliseconds. Ozone is a colorless gas that has an odor similar to smell of the air after a major thunderstorm.

Ozone has a greater disinfection effectiveness against bacteria and viruses compared to chlorination. In addition, the oxidizing properties can also reduce the concentration of iron, manganese, sulfur and reduce or eliminate taste and odor problems. Ozone oxides the iron, manganese, and sulfur in the water to form insoluble metal oxides or elemental sulfur. These insoluble particles are then removed by post-filtration. Organic particles and chemicals will be eliminated through either coagulation or chemical oxidation. Ozone is unstable, and it will degrade over a time frame ranging from a few seconds to 30 minutes. The rate of degradation is a function of water chemistry, pH and water temperature.

The formation of oxygen into ozone occurs with the use of energy. This process is carried out by an electric discharge field as in the CD-type ozone generators (corona discharge simulation of the lightning), or by ultraviolet radiation as in UV-type ozone generators (simulation of the ultraviolet rays from the sun). In addition to these commercial methods, ozone may also be made through electrolytic and chemical reactions. In general, an ozonation system includes passing dry, clean air through a high voltage electric discharge, i.e., corona discharge, which creates and ozone concentration of approximately 1% or 10,000 mg/L. In treating small quantities of waste, the UV ozonators are the most common, while large-scale systems use either corona discharge or other bulk ozone-producing methods.

The raw water is then passed through a venturi throat which creates a vacuum and pulls the ozone gas into the water or the air is then bubbled up through the water being treated. Since the ozone will react with metals to create insoluble metal oxides, post filtration is required.

Primary advantages of the technology are:

1. Ozone is effect over a wide pH range and rapidly reacts with bacteria, viruses, and protozoans and has stronger germicidal properties then chlorination. Has a very strong oxidizing power with a short reaction time.
2. The treatment process does not add chemicals to the water.
3. Ozone can eliminate a wide variety of inorganic, organic and microbiological problems and taste and odor problems. The microbiological agents include bacteria, viruses, and protozons.

The disadvantages are:

1. There is a higher equipment and operational cost and it may be more difficult to find professional proficient in ozone treatment and system maintenance.
2. Ozonation provides no germicidal or disinfection residual to inhibit or prevent regrowth.
3. Ozonation by-products are still being evaluated and it is possible that some by-products by be carcinogenic. These may include brominated by-products, aldehydes, ketones, and carboxylic acids. This is one reason that the post filtration system may include an activate carbon filter.
4. System may require pretreatment for hardness reduction or the additional of polyphosphate to prevent formation of carbonate scale.
5. Ozone is less soluble in water, compared to chlorine, and therefore special mixing techniques are needed.
6. Potential fire hazards and toxicity issues associated with ozone generation

7.4. Testing of Rainwater Quality

Harvested rainwater should be tested before drinking and periodically thereafter. Harvested rainwater should be tested both before and after treatment to ensure treatment is working. It is advisable to test water quarterly at a minimum, if used for drinking.

Before capturing rainwater samples for testing, contact the testing entity first to become informed of requirements for container type and cleanliness, sample volume, number of samples needed, and time constraints for return of the sample. For instance, for total coliform testing, water must usually be captured in a sterile container issued by the testing entity and returned within a maximum of 30 to 36 hours. Testing for pH, performed by commercial analytical laboratories must be done on site; other tests are less time-critical.

Regular chemical or microbiological testing of domestic rainwater tanks is not necessary and in most cases is not recommended. The following information should be considered before rainwater testing is undertaken.

7.4.1. Microbial hazards

Rainwater collected and stored in a domestic tank will contain a range of microorganisms. While most of these will be harmless, the safety of rainwater will depend on excluding or minimising the presence of pathogenic (disease-causing) organisms introduced into the water through faecal contamination.

Faecal contamination of rainwater tanks is often due to bird and small animal droppings washed in from the roof catchment area or directly into the tank itself in the case of poorly maintained tanks.

Other potential sources of contamination include dead animals and insects either in gutters or in the tank itself.

Rainwater collected in an underground tank that is not fully sealed or protected against ground run-off may become contaminated with microorganisms associated with human and animal excreta.

Accumulated sludge and sediment in a rainwater tank can harbour microorganisms capable of producing odours and affecting taste. These microorganisms are not hazardous to health but can result in unpleasant odours and tastes.

Off-odours from pipework and accumulated tank sediments are more common in hot or warm weather when intakes of water are absent or infrequent. It is not always practical to immediately clean a tank, particularly if it is the only source of water. In this case, chlorinating the tank can provide a temporary solution until cleaning can be undertaken.

Rainwater quality can change rapidly. Testing will only give an indication of the water quality on the particular day the sample was taken, not the overall or long-term water quality. Testing for specific pathogens can be expensive and is generally only warranted as part of an outbreak investigation. If there are strong concerns about water quality, chlorination of tank water could be a suitable alternative to testing. Should you decide to have your rainwater tested, *E. coli* is the indicator of choice for detection of faecal contamination. Tests for total coliforms or heterotrophic plate counts are not necessary.

7.4.2. Chemical hazards

Sources of chemical hazards can be divided into two types:

- those arising from off-site sources beyond the control of the owner/resident, including urban traffic, industrial emissions and poor agricultural practices
- those arising from on-site sources in the immediate vicinity of the tank and controllable by the owner/resident such as materials used in the construction of the roof, gutters, piping and tank and flues from wood heaters

Chemical testing is generally not required unless concerns exist about the quality of the water due to suspected impacts of major industrial or agricultural activities such as pesticide spray-drift, industrial emissions. If testing is being considered, the chemicals of concern need to be identified before testing, or large costs can be incurred with a limited likelihood of successful detection.

7.5. Water Quality Standards

No national or state standards exist currently for harvested rainwater quality. For those intending to harvest rainwater for potable use, the microbiological contaminants *E. coli, Cryptosporidium, Giardia lamblia,* total coliforms, legionella, fecal coliforms, and viruses, are probably of greatest concern, and rainwater should be tested to ensure that none of them are found. County health department and city building code staff should also be consulted concerning safe, sanitary operations and construction of rainwater harvesting systems.

Water sanitary inspections should be a focus of operational monitoring. This should include checking the cleanliness of the catchment area and storage, the structural integrity of the system and the physical quality of rainwater (turbidity, colour and smell). The level of pH should be monitored frequently in case of new concrete, ferrocement or masonry storage tanks.

Microbial quality of rainwater needs to be monitored as part of verification. Rainwater, like all water supplies, should be tested for E. coli or thermotolerant coliforms. The levels of lead, zinc or other heavy metals in rainwater should also be measured occasionally when it is in contact with metallic surfaces during collection or storage.

Management plans should document all procedures applied during normal operation as well as actions in the event of failures. In the case of rainwater harvesting, remedial actions will generally involve physical repair of faults and cleaning of catchment areas, filters or storage systems. Disinfection of rainwater should be practised when microbial contamination is detected or sanitary inspections indicate a likelihood of contamination. Disinfection with chlorine can make rainwater safe for drinking.

Independent surveillance is desirable for ensuring the quality, safety and acceptability of water supply based on rainwater. The principal focus of surveillance apart from verification of compliance should be geared more towards evaluation of hygienic practices in collection, storage and use of rainwater and needs for development and refinement for improving water safety.

7.6. Water Balance and System Sizing

Water Balance and System Sizing The basic rule for sizing any rainwater harvesting system is that the volume of water that can be captured and stored (the supply) must equal or exceed the volume of water used (the demand).

The variables of rainfall and water demand determine the relationship between required catchment area and storage capacity. In some cases, it may be necessary to increase catchment surface area by addition of a rain barn or outbuilding to capture enough rainwater to meet demand. Cistern capacity must be sufficient to store enough water to see the system and its users through the longest expected interval without rain.

The first decision in rainwater harvesting system design is the intended use of the water. If rainwater is to be used only for irrigation, a rough estimate of demand, supply, and storage capacity may be sufficient. On the other hand, if rainwater is intended to be the sole source of water for all indoor and outdoor domestic end uses, a more precise reckoning is necessary to ensure adequate supply.

In theory, approximately 0.62 gallons per square foot of collection surface per inch of rainfall can be collected. In practice, however, some rainwater is lost to first flush, evaporation, splash-out or overshoot from the gutters in hard rains, and possibly leaks. Rough collection surfaces are less efficient at conveying water, as water captured in pore spaces tends to be lost to evaporation.

Also impacting achievable efficiency is the inability of the system to capture all water during intense rainfall events. For instance, if the flow-through capacity of a filter-type roof washer is exceeded, spillage may occur. Additionally, after storage tanks are full, rainwater can be lost as overflow.

For planning purposes, therefore, these inherent inefficiencies of the system need to be factored into the water supply calculation. Most installers assume an efficiency of 75 percent to 90 percent. In most locations, rainfall occurs seasonally, requiring a storage capacity sufficient to store water collected during rainy times to last through the dry spells.

The collection surface is the "footprint" of the roof. In other words, regardless of the pitch of the roof, the effective collection surface is the area covered by collection surface (length times width of the roof from eave to eave and front to rear). Obviously if only one side of the structure is guttered, only the area drained by the gutters is used in the calculation.

7.6.1. Rainfall Distribution

To ensure a year-round water supply, the catchment area and storage capacity must be sized to meet water demand through the longest expected interval without rain. For instance, in West Texas, the historic longest span of continuous dry days has exceeded three months. If the rainwater harvesting system is intended to be the sole water source for a household, the designer must size the system to accommodate the longest anticipated time without rain, or otherwise plan for another water source, such as a well backup or hauled water. Also, rainfall from high-intensity, short duration rainfall events may be lost to overflow from storage tanks or splashout from the gutters. Although these intense rainfall events are considered part of the cumulative annual rainfall, the total available volume of such an event is rarely captured.

Another consideration is that most rainfall occurs seasonally; annual rainfall is not evenly distributed throughout the 12 months of the year. The monthly distribution of rainfall is an important factor to consider for sizing a system.

Two different estimators of monthly rainfall are commonly used: average rainfall and median rainfall. Average annual rainfall is calculated by taking the sum of historical rainfall and dividing by the number of years of recorded data. Median rainfall is the amount of rainfall that occurs in the midpoint of all historic rainfall totals for any given month. In other words, historically for the month in question, half of the time the rainfall was less than the median and half of the time rainfall was more than the median.

Median rainfall provides for a more conservative calculation of system sizing than average rainfall. The median value for rainfall is usually lower than the average value since large rainfall events tend to drive the average value higher. In other words, the sum of monthly medians is lower than the annual average due to the fact that the arithmetic average is skewed by high-intensity rainfall events. For planning purposes, median monthly rainfall can be used to estimate water availability to a reasonable degree of certainty.

7.6.2. Calculating Storage Capacity

Once the median or average potential for rainfall capture is known from rainfall data and catchment area, it will be necessary to calculate storage capacity. The decision of whether rainwater will be used for irrigation, potable and domestic use, or both, will dictate water demand, and therefore, capacity.

A simple method of roughly estimating storage capacity popular among professional installers is to size the storage capacity to meet quarterly demand. The system is sized to meet estimated demand for a three-month period without rain. Annual estimated demand is divided by four to yield necessary storage capacity using this approach. This approach, however, may result in a more expensive system due to higher storage costs.

If a rainwater harvesting system is to be the sole water supply, overbuilding ensures a safety margin. As with many things in life, it helps to hope for the best but plan for the worst. Even when budget constraints may not allow the user to install as much storage capacity as a sizing method may indicate, it is important to provide for an area where additional tanks or cisterns can be installed at a later date when finances permit.

7.6.3. Water Balance Method by Monthly Demand and Supply

One method of determining the feasibility of a proposed system is the monthly water balance method. This method of calculation is similar to maintaining a monthly checkbook balance. Starting with an assumed volume of water already in the tanks, the volume captured each month is added to the previous balance and the demand is subtracted. The initial volume of water in the tanks would be provided by hauling or capturing water prior to withdrawing water from the system.

Data and calculations can be entered on an electronic spreadsheet to enable the user to compare different variables of catchment area and storage. It is suggested that homeowners experiment with different variables of storage capacity and, if applicable, catchment surface to find individual levels of comfort and affordability for catchment size and storage capacity.

As mentioned above:

- catchment area and rainfall determine supply, and
- demand dictates required storage capacity.

A commitment to conserving water with water-saving fixtures, appliances, practices indoors, and low-water-use landscaping outdoors is an essential component of any rainwater harvesting system design. Not only is conservation good stewardship of natural resources, it also reduces the costs for storage capacity and related system components. If the amount of rainwater that can be captured – calculated from roof area and rainfall – is adequate or more than adequate to meet estimated demand, and meets the physical constraints of the building design, then storage capacity can be sized to meet estimated demand. If the monthly amount of water that can be captured, accounting for dry spells, is less than monthly estimated demand, then additional catchment area or supplemental supplies of water (such as groundwater from a well) will need to be considered.

In drier areas, no matter how large the storage capacity, catchment area may need to be increased with a rain barn or additional roof area to meet demand.

7.6.3.1. *Estimating demand*

A water-conserving household will use between 25 and 50 gallons per person per day. Households served previously by a water utility can read monthly demand from their meter or water bill to find monthly demand for purposes of building a new rainwater harvesting system. Divide the monthly total by the number of people in the house, and the days in the month to get a daily per capita demand number.

7.6.3.1.1. Indoor water conservation

Indoor domestic water conservation can be achieved by a combination of fixtures, appliances, and water conserving practices. The advantage of water-conserving appliances is that they require no change in household routine. Some water-conserving practices need user action, such as turning off the water while brushing teeth or shaving; washing vegetables in a pan rather than under a stream; washing only full loads of laundry and dishes; and keeping a pitcher of water in the refrigerator, rather than waiting for cold water to arrive from a faucet.

Water conservation appliances include:

— *Ultralow flush toilets (ULFTs).* Since 1993, only ULFTs with 1.6 gallons per flush may be sold in the United States. Older toilets should be replaced with the more efficient models. Some of the ULFTs require special early closing flappers to maintain their low-flow rates, so care should be taken in purchasing the correct replacement flapper for leaking toilets. If purchasing a new toilet, those that do not use early closure flappers are recommended. Dual-flush toilets (using less volume for liquid wastes) are also a good choice for a water-wise household.

— *Faucet aerators and efficient showerheads.* These fixtures are designed to use 2.2 gallons per minute at 60 psi, or 2.5 gpm at 80 psi. Studies have shown that most people feel comfortable at less than full flow rates, so using the new fixtures should provide you with an efficient and comfortable experience.

— *Hot water on demand.* These wallmounted units heat water just prior to use, eliminating the waste of waiting for hot water from the water heater while cold water is allowed to flow down the drain. Hot water loop systems keep hot water continuously circulating to achieve the same goal, but can use more energy. Another on demand unit heats water quickly only when activated by a pushbutton, rather than circulating water through a loop, saving both water and energy. A rebate from San Antonio Water System (SAWS) is available for installation of this type of on-demand circulation system.

— *Horizontal-axis (front-loading) clothes washers.* Because clothes are tumbled through a small volume of water in the bottom of the drum, this appliance can save up to half the water of a traditional clothes washer. It is also as much as 42 percent more energy efficient. A list of frontloading, horizontal-axis clothes washers is maintained by the Consortium for Energy Efficiency online.

7.5.3.1.2. Estimating outdoor water demand

Outdoor water demand peaks in hot, dry summer. In fact, as much as 60 percent of municipal water demand in the summer is attributable to irrigation. The water demands of a large turfgrass area almost always preclude the sole use of harvested rainwater for irrigation. Many urban dwellers capture rainwater for irrigation of vegetable and ornamental gardens. Because it is free of salts and minerals, rainwater promotes healthy plant growth. In urban areas, rainwater harvesters may reduce their water bill by substituting harvested rainwater for municipal water for garden irrigation.

For both the health of landscape plants and water use-efficiency, the best way to water plants is according to their needs. Signs of water stress include a gray blue tint to leaves, leaf rolling, and in the case of turfgrass, a footprint that does not spring back. Watering infrequently and deeply has been shown to promote plant health, waiting until plants need the water helps the water user to be sure that they are growing a healthy landscape.

For planning purposes, historical evapotranspiration can be used to project potential water demands. Evapotranspiration is the term for water use by plants, the combination of evaporation from the soil and transpiration from the plant leaves.

References

Lye D. "Health risks associated with consumption of untreated water from household roof catchment systems". *Journal of the American Water Resources Association* 38(5):1301-1306. 2002.

Macomber P. "Guidelines on rainwater catchment systems for Hawaii". College of Tropical Agriculture and Human Resources, University of Hawaii at Manoa. 2001.

Thomas PR, Grenne GR. 1993. "Rainwater quality from different roof catchments". *Water Science Technology* (28):290-99. EPA, www.epa.gov/safewater/mcl.html

8 Rainwater Harvesting and Ecosystem Services

Rainfall and soil water are fundamental parts of all terrestrial and aquatic ecosystems which supplies goods and services for human well-being. Availability and quality of water determines ecosystem productivity, both for agricultural and natural systems. There is increasing demand on water resources for development whilst maintaining healthy ecosystems, which put water resources under pressure. Ecosystem services suffer when rain and soil water becomes scarce due to changes from wet to dry seasons, or during within-seasonal droughts. Climate change, demand for development and already deteriorating state of ecosystems add to these pressures so that future challenges to sustain our ecosystems are escalating.

8.1. Effect of Rainwater Harvesting on Ecosystems

Rainwater harvesting is often an intervention intended to augment the provisioning services of the environment for human well-being. Provisioning services, as defined in the Millennium Ecosystem Assessment, include environmental services such as improved and safe water supplies, or increased crop production. A closer analysis shows that rainwater harvesting often has many more impacts, both positive and negative on ecosystem services, and extending to regulating, cultural and supporting services.

Water is essential for all living beings, for consumptive use. Plants and vegetation are by far the largest water consumers, but they also provide direct livelihood and economic returns to humans as food, fodder, fibres and timber, in addition to products for pharmaceutical use, diverse genetic resources and fresh water.. Abstraction of water for human use is circa 3,600 km^3, or 25 % of renewable freshwater flows annually. These abstractions mainly provide irrigation water (70%) to increase crop production. Use of water for drinking, and public, commercial and other societal needs is essential but relatively minor in quantity, and much is returned to landscape, often through waste water systems.

Table 8.1. Ecosystems functions and the effect of rainwater harvesting

Ecosystem services	*Effect of rainwater harvesting intervention*
Provisioning	can increase crop productivity, food supply and income
	can increase water and fodder for livestock and poultry
	can increase rainfall infiltration, thus recharging shallow groundwater sources and base flow in rivers
	can regenerate landscapes increasing biomass, food, fodder, fibre and wood for human consumption improves productive habitats, and increases species diversity in flora and fauna
Regulating	can affect the temporal distribution of water in landscape
	reduces fast flows and reduces incidences of flooding
	reduces soil erosion
	can provide habitat for harmful vector diseases
	bridges water supply in droughts and dry spells
Cultural	rainwater harvesting and storage of water can support spiritual, religious and aesthetic values
	creates green oasis/mosaic landscape which has aesthetic value
Supporting	can enhance the primary productivity in landscape
	can help support nutrient flows in landscape, including water purification

Rainwater harvesting is a way of increasing the provisioning capacity at a specific location. Many rainwater harvesting interventions to date are primarily to increase crop/fodder/food/timber production, or to provide domestic/public/commercial supplies of water.

8.1.1. Regulating Ecosystem Services

The regulating services are in addition to the supporting services, discussed below, and are essential for human well-being as they control the type and provisioning capacity of ecosystems in specific locations. Water flows across the landscape play a role in a range of regulation services as water is primarily involved in many of them. The primary roles of the presence (or absence) of water are in erosion control, climatic control, pest and disease control (through habitat regulation), water quality control and control of natural hazards.

Implementation of rainwater harvesting interventions may affect the regulating services of the landscape as the landscape water flows change. As mentioned earlier, soil conservation measures to reduce soil erosion also act as *in situ* rainwater harvesting measures. *Ex situ* rainwater harvesting and storage in the urban and rural landscape affects flooding and flow duration over seasons. Increasing the numbers of ponds and dams storing harvested rainwater in the landscape may increase the incidence of malaria, but if covered, or if water is stored underground, this may not impact incidence of malaria in the specific location.

8.1.2. Cultural Ecosystem Services and Rainwater Harvesting

Water has strong cultural and religious values. These values are critical for human spiritual well-being, and are recognised as having an essential role in societal interactions, once primary resources are provided. Water also has an aesthetic value, enhancing garden and ornamental plant growth, and providing green "oases," for example, in urban areas.

Increasing access to water through rainwater harvesting in a community or household may act to enhance the access and ability to carry out religious and spiritual rituals. It can also increase the aesthetic use of water. At the landscape scale, water features are often protected and given specific values and protection by the local community.

8.1.3. Supporting Ecosystem Services

The supporting services pre-determine the conditions for all other services. Water flows play an essential role as a medium for the transport of nutrients and contaminants, in the shaping of soils, and in photosynthesis. Together with soil conditions and climate conditions the water balance will determine the net primary production level at a given location.

Rainwater harvesting will not primarily affect these supporting services. Indirectly, soil formation may change from a natural course as *in situ* management interventions are implemented. Also leakage of nutrients may change, but indirectly, mainly due to changing from natural land use patterns to agricultural uses rather than from implementing *in situ* rainwater harvesting in the fields.

To conclude, rainwater harvesting is often implemented to improve local provisioning capacity by ecosystems for human well-being. However, as the landscape water balance is affected by increased rainwater harvesting, other services, in particular regulating services related to water abundance and availability, can be affected. Cultural services can be either negatively (if resources are diminished due to rainwater harvesting) or positively, depending on the local context.

8.2. Rainwater Harvesting and Landwater Flows

Rainfall is the main source of freshwater in all landbased ecosystems, whether natural or managed by humans. From arid deserts to the humid tropical rainforests, the flow of water through the ecosystem shapes the characteristic fauna and flora as well as the soil systems. The land surfaces globally receive 113,000 km^3 of rainfall. Of this, approximately 41,000 km^3 (36%) is manifested as surface runoff in the liquid phase—the so-called 'blue water' of rivers, streams and lakes. The remaining amount, 64%, of the rainfall, is evaporated through vegetation, from soil surfaces and from water surfaces within the landscape.

Rainwater harvesting is principally the management of these two partitioning points in the water flow. At the local scale, such as a farm field, the flows partition the incoming rainfall at the soil surface, either infiltrating the water into the soil or diverting the water as surface runoff. Within the soil, the second partitioning point is at the plant roots where water is either taken up by the vegetation, or contributes to the recharge of shallow or deep groundwater. Depending on the soil surface conditions, infiltration can range from 100% in a well managed agricultural soil, to 100% runoff from a paved road or rooftop. The second partitioning point can be managed, indirectly, through planting different plant species, improving crop uptake capacity, modifying plant root depths, and altering soil management practices in agriculture, through enhancing/ depleting soil health, including soil organic matter content. *In situ* rainwater harvesting interventions address both partitioning points at the field scale. Many soil management practises, such as soil conservation measures that enhance the soil infiltration capacity and soil moisture storage, can alter the partitioning process. Practises that enhance root water uptake for crop growth act as an '*in situ* rainwater harvesting' intervention. *Ex situ* rainwater harvesting, in contrast, alters the partitioning process at the local field scale.

At the landscape scale (or meso-scale, 1 km^2-10,000 km^2), rainfall partitioning and flow paths are the same as at the field scale, but the quantities cannot simply be aggregated from field scale to landscape scale, as water often re-distributes itself within the field, and/or along a slope gradient. When rainwater harvesting interventions are implemented, the partitioning is changed. One change is to increase infiltration and storage of water in the soil. This has short term advantages as it slows the flow of water, which reduces soil erosion, minimises flooding and limits damage to built structures due to storm water flows. A longer term advantage is an effect of the slower flows of water within the landscape. The longer residence times enable water to be accessed during dry periods, and used for productive purposes, including human consumption, livestock watering and increased crop and vegetation growth.

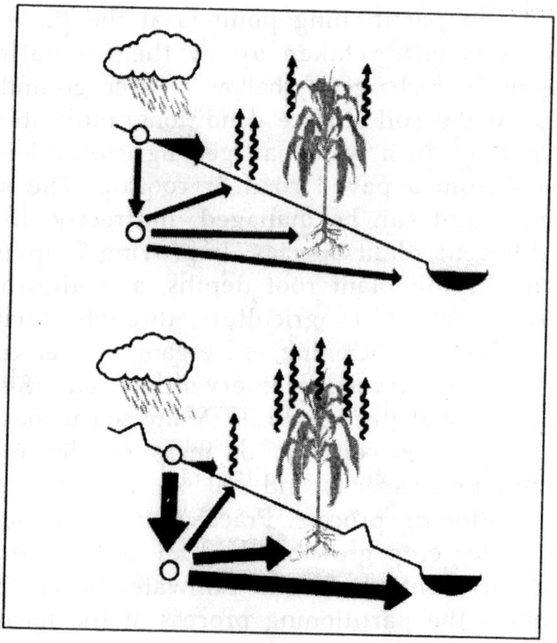

Figure 8.1. Landscape water balance flows a) without rainwater harvesting, and b) an example of flow paths with rain water harvesting interventions with water partitioning points at the soil surface and in the soil. Rainwater harvesting is principally about managing water partitioning in these points.

While rainwater harvesting can increase crop and other vegetation productivity through improved water access, reducing soil erosion and incidences of flooding downstream, harvested rainfall may increase depletion of downstream users' access to water conveyed downstream as surface runoff or downgradient as groundwater. At a certain point, if the consumptive use of water resources such as for crop or other vegetation growth is complete, the loss of downstream access to the water may be severe and irreversible.

Increasingly, it is recognised that the multiple interventions by humans on ecosystems sometimes create unexpected and irretrievable changes in the services provided. These unexpected changes are often referred to as 'tipping points', where an ecosystem or its services shift from one production regime to another. In water resource management such tipping points have been experienced in watershed and river basins subject to excessive consumptive and re-allocation of water resources. An example is the Aral Sea, which due to irrigation water outtake, is permanently damaged with concomitant reductions in ecosystem services generated. At a smaller scale, excessive erosion can alter a field to an unproductive state, and, with a single event, possibly irretrievably damage the field through land subsidence or a landslide. With increasing interventions to abstract water, communities and resource managers should be aware that interventions at different scales can feedback unexpectedly, and erode ecosystem services permanently. On the other hand, efficient and productive water and land usage, for example through many smallscale rainwater harvesting interventions, has shown positive change, where interventions have resulted in increased opportunity and productivity of ecosystem services.

Further interventions may affect the landscape water flows so it is impossible to restore downstream or downgradient access, i.e., the water balance undergoes a regime shift. Such regime shifts include altering the timing of delivery of surface runoff, for example when deforestation or afforestation occurs, or when irrigated agriculture affects groundwater levels and/or water quality. Shifts in flow regimes are difficult to remediate. To date, there is limited synthesised evidence to document the impact of rainwater harvesting on downstream water flows.

8.3. Rainwater and Human Ecosystem

Water is an essential commodity for all living beings: for direct consumption to sustain life and health, for indirect consumption through water required to grow food, fodder and fibres, and for maintenance of the range of ecosystem services needed to support and sustain economic and social activities. Water's role in promoting human well-being has been variously defined. Four key areas stand out as particularly important when linking water with improvements of human well-being:

— *Water and health:* domestic water supplies for human consumption, hygiene and sanitation;

— *Water and basic provisioning*: water for producing food, fodder and fibres;

— *Water and livelihoods:* water to support rural livelihoods and sustain economic activity; and

— *Water and vulnerability:* water as a component in natural disasters and disaster mitigation.

A globally accepted set of indicators of human wellbeing are the Millennium Development Goals and associated targets, which were developed and agreed in 2000. Rainwater harvesting can play both a direct and indirect role in the achievement of many of these goals, particularly in the area of basic human needs and health.

A more comprehensive view of human well-being is taken by the Millennium Ecosystem Assessment in which human well-being is not only a result of good health and adequate basic provision of food, shelter and other material necessities, but also related to freedom of choice and action, security and the need for good social relations. In this context several cases of rainwater harvesting, especially as an element in watershed management, can play a significant role, especially for social relations, where water management has long been a unifying factor. There is increasing evidence that watershed management with rainwater harvesting has strengthened social capital which in turn can have a significant impact on development of other ecosystem services for human well-being.

8.4. Increasing Pressure on Water and Ecosystem Services

Several MDGs are closely related to water for health and sanitation. The MDG Target 7C aims to halve water supply deficits, presenting a formidable challenge for investment and social and technical alignment. However, the amounts of water necessary to reduce water supply deficits are in many cases available. In addition, the use of domestic water is not necessarily consumptive, as the water can be cleaned and re-used. Quantifying minimum water requirements to meet basic human needs has resulted in vastly disparate estimates. Annual per capita water needs range between 18 m^3 and 49 m^3, suggesting that approximately 0.1 to 0.3 km^3 is required for basic water consumption, sanitation and societal uses by the global population. It is important to note that water for domestic, public and commercial use in many cases is returned to stream flow locally. Return flows are reduced by consumptive losses, and often result in diminished water quality, increased health risks amongst downstream users and degraded habitats.

Relatively larger amounts of water are used to generate the ecosystem services needed to ensure provisioning of basic supplies of food, fodder and fibres. Just to meet the food requirements of a balanced diet, approximately 1,300-1,800 m^3 of water per person are consumed per year. This translates to 8,800-12,200 km^3 for a world population of 6.7 billion in 2008/2009. The water used for food production, whether irrigated or rainfed, is consumptive; i.e., at a local site, water will be incorporated into foodstuffs, evaporated from the land surface or otherwise non-retrievable for further use downstream. In comparison with amounts of water needed for domestic, public and commercial purposes, the projected needs for additional water to meet MDG target on hunger suggest additional withdrawals of water for both rainfed and irrigated agriculture to meet the target through 2015.

Today, rainfed and irrigated agriculture appropriate 7,700 km^3 of freshwater globally to provide food. Of this, approximately 2,600 km^3 is direct withdrawals for irrigation purposes. To meet the MDG Hunger goal, an additional volume of 1,850 to 2,200 km^3 of water needs to be appropriated annually, based upon current agricultural practises and assuming balanced diets. To feed all a reasonable diet by 2050 may require almost doubling of today's water resources. With renewable accessible freshwater globally limited to 12,500 km^3, it is a great challenge facing humanity. The consumptive use of water for crops and vegetation to provide other biomass goods such as timber, fibres for clothing, wood for energy etc. is not included in the above numbers.

A third dimension is the sustainable management of resources. This is mainly addressed in MDG 7. This target can be interpreted as seeking to ensure sustainable use and safeguarding of water resources. Such safeguarding could include the management of water for other uses, for example, for ecosystem services, including provision of minimal environmental flows necessary for maintenance of aquatic organisms and their habitats. Accounting for the provision of minimum environmental flows in major river basins suggests that water stress is even more imminent than when estimated based on renewable water resources solely for human use. These estimates suggest that, already, 1.1 billion people are living in severely water stressed basins, and an additional 700 million people live in moderately stressed river basins. Clearly, further consumptive use of water or increased pollution may seriously affect ecosystem health, as well as human well-being and potential for development.

8.5. Typology of Rainwater Harvesting Systems

Definition and typology of rainwater harvesting systems Rainwater harvesting consists of a wide range of technologies used to collect, store and provide water with the particular aim of meeting demand for water by humans and/or human activities. These technologies can be divided into two main areas depending on source of water collected; namely, the *in situ* and the *ex situ* types of rainwater harvesting, respectively. In essence, *in situ* rainwater harvesting technologies are soil management strategies that enhance rainfall infiltration and reduce surface runoff. The *in situ* systems have a relatively small rainwater harvesting catchment typically no greater than 5-10 m from point of water infiltration into the soil. The rainwater capture area is within the field where the crop is grown.

In situ systems are also characterised by the soil being the storage medium for the water. This has two principal effects. Firstly, it is difficult to control out-take of the water over time. Normally soil moisture storage for crop uptake is 5-60 days, depending on vegetation type, root depth and temperatures in soil and overlying atmosphere. Secondly, the out-take in space is determined by the soil medium characteristics, including slope. Due to gradients and sub-surface conditions, the harvested water can act as recharge for more distant water sources in the landscape, including groundwater, natural water ways and wetlands, and shallow wells.

The *in situ* rainwater harvesting systems are often identical to a range of soil conservation measures, such as terracing, pitting, conservation tillage practices, commonly implemented to counter soil erosion. Thus, harvesting rainwater by increasing soil infiltration using *in situ* technologies also counteracts soil loss from the farmed fields or forested areas. *In situ* rainwater harvesting often serves primarily to recharge soil water for crop and other vegetation growth in the landscape. The water can also be used for other purposes, including livestock and domestic supplies if it serves to recharge shallow groundwater aquifers and/or supply other water flows in the landscape.

The *ex situ* systems are defined as systems which have rainwater harvesting capture areas external to the point of water storage. The rainwater capture area varies from being a natural soil surface with a limited infiltration capacity, to an artificial surface with low or no infiltration capacity. Commonly used impermeable surfaces are rooftops, roads and pavements, which can generate substantial amounts of water and which can be fairly easily collected and stored for different uses.

As the storage systems of *ex situ* systems often are wells, dams, ponds or cisterns, water can be abstracted easily for multiple uses including for crops and other vegetation as irrigation water, or for domestic, public and commercial uses through centralised or decentralised distribution systems. By collecting and storing water in dams, tanks, and cisterns the storage time is more dependent on the size of capture area, size of storage unit and rate of outtake rather than residence time and flow gradient through the soil.

The wide variety of rainwater harvesting technologies and end uses of the water also indicates the dynamic and flexible dimensions of rainwater harvesting systems. They also reflect the multiple end uses of the water collected for our benefit, including agriculture and landscape management, domestic, public and commercial water supply, as well as livestock watering, aquaculture and maintaining aesthetic values.

8.6. Implementation of Rainwater Harvesting Systems

There is much historical evidence of rainwater harvesting being an important factor in community development since the beginning of human settlements. Many cultures have developed their societies with the primary management of water resources as a corner stone, developing more sophisticated ways of supplying water both for consumption and agriculture. Rainwater harvesting structures using cisterns are dated as early as 3000 BC in the Middle East. A more in-depth description of ancient rainwater harvesting in India has been summarised by the Centre for Science and Environment, India.

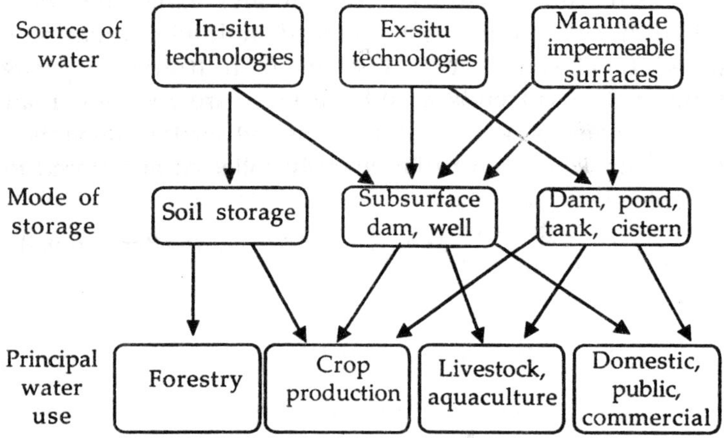

Figure 8.2. Schematic of Rainwater Harvesting Technologies based on Source of Water and Water Storage Type

At the global level, there is no comprehensive assessment of the extent of implementation of rainwater harvesting technologies for specific uses. Nor is there any summarised information on how much land is currently under any type of *in situ* rainwater harvesting. For the specific application of conservation tillage, as no tillage agriculture, national statistics have been aggregated by Hobbs *et al*. Their information suggests that, globally, only a small fraction of the land surface, amounting to about 95 million hectares, is currently under conservation or no–till agriculture.

Unfortunately, the information on irrigation cannot directly be associated with rainwater harvesting systems for irrigation purposes as it differentiates between surface water and groundwater, which does not allow the separation of shallow groundwater from deep groundwater, nor surface water withdrawn from 'blue' water sources from smaller scale systems.

The recent assessment of irrigated and rainfed land, completed in the Comprehensive Assessment of Water Management in Agriculture, also did not differentiate areas under rainwater harvested water supply from areas under other types of water supply for irrigation. This lack of global information on where and how much rainwater harvesting is currently in use makes it impossible to say how many people actually benefit from rainwater harvesting today. It also becomes challenging to summarise the global and/or regional benefits and costs in specific locations, countries or regions of rainwater harvesting for human well-being or ecosystem impacts arising from rainwater harvesting.

8.7. Rainwater Harvesting for Watershed Ecosystems

Watershed management and development refers to the conservation, regeneration and the judicious use of the natural (land, water, plants, and animals) and human habitat within a shared ecosystem (geological hydrological-aquatic and ecological) located within a common drainage system. Over the years, watershed management has come to be seen as the initiation of rural development processes in arid and semi arid areas, in particular in rainfed ecosystems – combining projects for ecological sustainability with those for socioeconomic development.

Theoretically, it attempts to integrate sectors such as water management, agriculture, forestry, wasteland development, off-farm livelihood development, etc., and to establish a foundation for rural development. The approach aims to be flexible enough to be adapted to varying sociological, hydrological and ecological conditions. Apart from the purely environmental concerns, i.e., restoring ecosystem functions, the watershed framework often focuses on livelihood improvements, poverty alleviation and a general increase in human well-being.

Watershed management is a strategy which responds to the challenges posed by a rainfed agro-ecosystem and human demands. Typically these challenges include water scarcity, rapid depletion of the ground water table and fragile ecosystems, land degradation due to soil erosion by wind and water, low rainwater use efficiency, high population pressure, acute fodder shortage and poor livestock productivity, mismanagement of water sources, and lack of assured and remunerative livelihood opportunities. Therefore, the watershed management approach seeks to ensure human well-being and progress toward sustainable development through improved ecosystem services—including food, fresh water, fuel wood, and fiber.

Changes in availability of all these ecosystem services can profoundly affect aspects of human well-being – ranging from the rate of economic growth and level of health and livelihood security to the prevalence and persistence of poverty. The framework of watershed management acknowledges the dynamic interrelationship between people and ecosystems. To bring about a positive change in the ecosystem services of the local habitat, the watershed management approach deals with people and ecosystem in a holistic and inter-disciplinary way.

The water management component of watershed management in rainfed areas largely depends on rainwater to initiate the local development processes. Thus, the aim of this chapter is to highlight some of the critical issues facing rainwater harvesting in watershed management, against the backdrop of human and ecosystem well-being.

Watersheds consist of a complex pattern of various ecosystems (forests, farmland, wetlands, soils, etc) which provide a number of important goods and services for human well-being. Examples are ample and safe water supply from rivers and groundwater, crops, fish, fuel and fibres, as well as flood and erosion control. Rainwater is, by itself, an important input factor for healthy and productive ecosystems.

Rainwater harvesting in the context of a watershed means collecting runoff from within a watershed area, storing it, and employing it for different purposes. Runoff collection is generally distinguished as in situ management, when the water is collected within the area of harvesting, and *ex situ* when it is diverted outside of the harvesting area. The storage is of crucial importance: for in situ rainwater harvesting the soil acts as the storage, whereas for *ex situ* rainwater harvesting the reservoir can be natural or artificial, where natural generally means groundwater recharge, and artificial means surface/subsurface tanks and small dams. The differentiation between the two is often minor, as water collection structures are generally placed in a systematic relation with each other; hence, the runoff from certain structures may be a source of recharge for others. For example, the construction of anicuts (small dams) at frequent intervals in seasonal rivers leads to increased groundwater recharge. Rainwater harvesting in a watershed context has a role and an impact on several aspects of ecosystems and human well-being. This section will present a few of them, through examples and case studies.

8.7.1. Impacts on Downstream Flows

Amongst the proponents of rainwater harvesting, the argument in favour of its potential to drought-proof India has developed so far as to prove that, if half of rainfall is captured, every village in India can meet its own domestic water needs. The strategy for drought proofing would be to ensure that every village captures all of the runoff from the rain falling over its entire land and the associated government revenue and forest lands, especially during years when the rainfall is normal, and stores it in tanks or ponds or uses it to recharge depleted groundwater reserves. It would then have enough water in its tanks or in its wells to cultivate substantial lands with water-saving crops like millet and maize. Although detractors highlight the variability of rainfall and potential effect of heavy harvesting on downstream water resources during drought years, the resonance of this argument is strong.

Rainfall can cover basic human needs in dry areas in a decentralised and sustainable way and thus reduce pressures on pressures of fragile groundwater reserves. These estimates prove that the potential of rainwater harvesting is large and that there is little reason why a village, region, or a country has to experience water problems, if they have land and rains. However, one of the conditions of sustainable watershed management is to recognise so-called negative externalities.

In this case the negative externality would be the effects of rainwater harvesting on downstream water availability. Runoff out of the watershed may be considered as a waste from a local point of view, but it may be a key resource for surface withdrawals or recharge of groundwater for downstream users. For example, the Sardar Patel Participatory Water Conservation programme was launched by the government of Gujarat in Saurashtra and north Gujarat in 1999, and involved the building of check dams in local streams, and nallas (drains). As the government of India officially claimed in 2007, nearly 54,000 check dams were built in Saurashtra and north Gujarat with the involvement of local communities. However, some caution has been raised, as this large and fast expansion of water harvesting potentially can affect the ecology of Saurashtra region.

8.7.2. Decentralised Approach

Given the fact that rainfall is unevenly distributed between years, as well as within rainy seasons, storing rainwater is a key component of water management. The water can be stored in storages of different construction and dimensions; for example, large reservoirs with large catchments and small tanks and ponds with small catchments, or use of natural or artificial groundwater recharge to store water in the soil.

There is evidence to show that village-scale rainwater harvesting will yield much more water for consumptive use than large or medium dams, making the latter a wasteful way of providing water, especially in dry areas. In the Negev desert where rainfall is only 105 mm annually, it was found that more water is collected if the land is broken up into many small catchments, as opposed to a single large catchment. This is because small watersheds provide an amount of harvested water per hectare which is much higher than that collected over large watersheds, as evaporation and loss of water from small puddles and depressions is avoided. As much as 75% of the water that could be collected in a small catchment is lost at the larger scale. It is important to recognise that the nonharvested water does not necessarily go to waste, as it is returned to the water cycle from the landscape.

Several other studies conducted by the Central Soil and Water Conservation Research Institute in Agra, Bellary and Kota, and another study conducted in the high rainfall region of Shillong, have all found that smaller watersheds yield higher amounts of water per hectare of catchment area. To put it simply, this means that in a drought-prone area where water is scarce, 10 tiny dams, each with a catchment of 1 ha, will collect more water than one larger dam with a catchment of 10 ha. However, critics have suggested that the benefits of smaller rainwater harvesting systems versus large scale downstream implementation is mostly an effect of different scale and project implementation, and lack of consideration of (negative) externalities. There is scientific evidence that even withdrawal of water by rainwater harvesting can have depleting effects, if the water is for consumptive uses such as irrigation. Evapotranspiration of plants (crops, trees, other vegetation) is an absolute loss of water, which potentially can affect downstream flows of water if used upstream excessively.

8.7.3. Increasing Infiltration and Groundwater Recharge

Groundwater recharge in watershed management can be induced through different structures; for instance, through dug shallow wells and percolation tanks. The estimated number of dug shallow wells in varying formations and situations in Rajasthan is about 83,000 wells, with potential new nadis (village ponds) estimated at 14,500.

The existing nadis and the ones to be built may contribute 360-680 million m^3 of groundwater replenishment annually. Percolation tanks alternatively are another recharge structure which is generally constructed on small streams and used for collecting the surface runoff. Under favourable hydro-geological conditions, percolation rates may be increased by constructing recharge (intake) wells within percolation tanks. According to studies conducted on artificial recharge, the percolation tanks constructed in hard rock and alluvial formations in the Pali district of Rajasthan had a percolation rate of 14 to 52 mm/day.

Percolation accounted for 65-89% of the loss whereas the evaporation loss was only 11-35% of the stored water. The results also indicated that the tanks in a hard rock area contained water for 3-4 months after the receding of the monsoon. Percolation tanks have been of greater benefit in recharging groundwater in the neighbouring Gujarat state. There is a huge potential to adopt this technology in western Rajasthan as well, where groundwater depletion rates are very high. Thus, percolation tanks hold great promise for drought mitigation in regions having impermeable strata beneath a sandy profile, with limited water holding capacity but high percolation rates.

8.7.4. Reducing Soil Erosion

Rainfed areas are also confronted with problems of land degradation through soil erosion. Watershed management interventions through water harvesting are often synonymous with soil and water conservation. They act both to harvest rainfall and to conserve soil and water, as a mean of increasing farm productivity. The available evidence reveals that soil loss is reduced by about 0.82 tons per ha per year due to interventions in the watershed in India. The consequence of these soil conservation activities also is the reduction of siltation of downstream tanks and reservoirs that in turn reduce the need for maintenance. An example is provided by a comprehensive assessment of the Rajasamadhiyala watershed, Gujarat, India, conducted to assess the on-site impact of a watershed management programme as well as off-site impacts on two downstream watersheds. Inspection of the 40 year old check dam in the downstream portion of the Rajasamadhiyala watershed, showed that, two years after the check dams construction upstream, the check dam downstream was completely free from siltation whereas previously it had silted up every 2 years.

8.7.5. Intensification of Crop Production

Reduction of surface runoff was used to augment both surface and shallow groundwater reserves through in situ rainwater harvesting interventions. This had a direct benefit by expanding the irrigated area and increasing cropping intensity. On average, the irrigated area increased by 34%, while the cropping intensity increased by 64%. Such an impressive increase in the cropping intensity was not achieved in many surface irrigated areas in the country.

Action for Social Advancement's (ASA) work in Madhya Pradesh, India, provides an example of how the increased volume of rainfall infiltration and surface storage has resulted in additional irrigated area, contributing to increased crop output as well as cash crop production. The improved water availability in the soil, and irrigation supply, has enabled farmers to grow a second crop during the winter season, after the usual monsoon season.

The local cropping pattern has changed, and at present the farmers have started growing wheat during the winter, and rice and soybeans during the monsoon. As part of land development activities, several farmers have built small field bunds to retain water in the fields that are flooded during the monsoon to grow a rice crop, for wheat production during the subsequent dry season. Improving food security and economic security Rainwater harvesting can be instrumental to decentralised water supplies and local food security.

Local food security is as important as national food security. It has been proven that the overall increase in crop output, mainly from the second crop, and from the establishment of homestead (kitchen) gardens, has had an impact upon the amount of food available for domestic consumption. When rainwater harvesting at the household or community level enables rainfed farms to access a source of supplementary irrigation, the economic security also improves.

According to farmers in the ASA implementation area in Madhya Pradesh, the visible signs of improved economic security are increased incomes from the sale of marketable agricultural surpluses, that typically has led to a reduction in dependency and debt, to a decrease in the reliance on moneylenders, and to an increase in savings and investment in new assets (primarily agriculture related assets) or improvement in existing assets.

8.8. Rainwater Harvesting and Agroecosystems

Development of agriculture is the main landuse change that has affected and depleted many ecosystem services in favour of increased agricultural biomass production. The soil and climate sets the parameters for agro-ecosystems. Increasingly, the value of a healthy soil system is recognised as a key ecosystem service in sustaining agro-ecosystem production. Water flows in the soil system depend upon two principal factors; infiltration of rainfall, and water holding capacity of the soil. Rainwater harvesting for crops is therefore closely related to soil system management; namely, the actions taken to improve infiltration into the soil and to increase water holding capacity and fertility functions in the soil.

In the water management community, much attention has been devoted to irrigated agriculture, since it appears to be the major consumer of water, when compared to water requirements for domestic and industrial purposes. However, much less attention has been paid by water managers and investment institutions to the issues of rainfed agriculture. The distinctive features of rainfed agriculture in developing countries are that both productivity improvement and expansion have been slower in relation to irrigated agriculture. But as Pretty and Hine suggest, there is a 100% yield increase potential in rainfed agriculture in developing countries, compared to only 10% for irrigated crops.

Rainfed agriculture produces, and will continue to produce, the bulk of the world's food. It is practised on 80% of the world's agricultural land area, and generates 65-70% of the world's staple foods, but it also produces the most food for poor communities in developing areas. In Sub-Saharan Africa more than 95% of the farmland is rainfed, while almost 90% in Latin America, 60% in South Asia, 65% in East Asia and 75% in the Near East and North Africa are rainfed. In India, 60% of water use in agriculture originates from directly infiltrated rainfall.

Low and variable productivity is the major cause of poverty for 70% of the world's poor inhabiting these lands. At the same time there is growing evidence that agriculture continues to play key role in economic development and poverty reduction in the rainfed regions. Increased effort is needed to upgrade rainfed systems, from the point of view of improving soil water capacity and fertility. More efficient rainwater harvesting systems have a great role to play, especially in developing countries struggling to provide water and affordable food.

8.8.1. Rainfed Farming Systems

Yield gap analyses in tropical semi-arid and sub-humid areas found farmers' yield a factor of 2-4 times lower than optimally achievable yields, for major rainfed crops. Grain yields fluctuate around 1-2 t/ha, compared to optimal yields of over 4-5 t/ha. The large yield gap between attainable yields and farmers' practice, as well as between attainable and potential yields, shows that there is a large potential to improve yields in rainfed agriculture that remains to be tapped.

Rainfall is the crucial input factor in the rainfed production system. Its variation and uncertainty is high in areas of low rainfall and a major cause of low productivity and heightened distress among farmers. The last decade, in particular, has witnessed serious distress, even amongst the more enterprising, small and marginal farmers in the rainfed regions. They opted to replace traditional low value cereals with high value ones, and introduce intensive crops through borrowing but with little success. Adverse meteorological conditions, long dry spells and droughts caused extended moisture stress periods for crops, livestock and people. Such situations occur over large parts of poor countries in Asia and sub-Saharan Africa.

Limited food productivity and poverty at the household level is a major contributing factor in the further degradation of ecosystems, including deforestation, excessive abstraction of biomass, and possibly landscape habitat destruction, with biodiversity loss as a result. In particular, degraded soils with low productivity can send the relationship between the ecosystem and human well-being into a downward spiral, with diminishing yields, affecting farmers' livelihoods, and reduced capacity to restore and enhance the soil system's health to a more productive state.

8.8.1.1. Constraints of Rainfed Agriculture Systems

In the most arid zones (< 300 mm/annum), absolute water scarcity constitutes the major limiting factor in water provision. But in the vast semi-arid and dry sub-humid tropical regions, total seasonal rainfall is generally adequate to meet most needs and also to significantly improve agricultural water productivity, if it were evenly distributed. However, dry spells (or monsoon breaks), with little or no rainfall, occur in most cropping seasons during critical stages of plant growth. Soil moisture storage reaches critical limits and causes crop damage, or even failure. Most rural poor in Asia, Africa and Latin America experience water scarcity for agriculture, and consequentially poor yields and compromised livelihoods, due to the lack of public, private and individual investment in the provision of even small scale water infrastructure. Here adaptation to rainfall variability is the greatest water challenge.

Therefore, local harvesting of a small portion of the rainwater in wet periods, utilising the same for supplemental/protective irrigation during devastating dry spells, offers a promising solution in the fragile, rainfed regions of the world. As total rainfall is spread over a few rainfall events of high intensity (about 100 hours in whole season in semi-arid regions), in most rainfed regions in Asia and Africa much is lost to runoff and evapotranspiration. It is important to capture and convert a part of this into more productive use. The storm runoff may either be diverted directly and spread on the fields, or collected in inexpensive water storage systems.

Water harvesting techniques may be catchment systems, collecting runoff from a larger area. They include runoff farming, which involves collecting runoff from the hillsides and delivering it onto plain areas, and floodwater harvesting within a streambed using barriers to divert stream flow onto an adjacent area, thus increasing infiltration of water into the soil. Microcatchment water harvesting methods are those in which the catchment area and the cropped area are distinct, but adjacent to each other. Establishing catchment systems often necessitates ecosystem rehabilitation and conservation, in order to secure the runoff.

At the farm level, rooftop runoff collection may be successfully used for gardening. At the field level, in situ water harvesting methods focus on the storage capacity of the land surface and the soil. They include conservation tillage, field embankments, trenches, half-moon terraces and field terracing. Thus, the upper layers of the soil form an important part of the ecosystem, which act as a natural storage for rainfall. The infiltrated rainfall has an important role in supporting soil fertility cycles as well as micro-flora and fauna in the soil.

Generally, the amount of water made available through rainwater harvesting is limited and has to be used most judiciously to alleviate water stress during critical stages of crop growth. Supplemental irrigation is a key strategy, so far underutilised, in unlocking rainfed yield potentials. It is used on crops that can be grown using rainfall alone, but provides a limited amount of water during times of low/no rainfall. The use of supplemental irrigation, to bridge dry spells, has the potential to substantially increase yields by more than 100%. The existing evidence indicates that supplemental irrigation, ranging from 50-200 mm/season (500-2000 m^3/ha), is sufficient to alleviate yield-reducing dry spells in most years, and thereby stabilise and optimise yield levels. In addition, supplemental irrigation systems have shown a further gain through improved water productivity; i.e., through gains in absolute consumption of water for the same production of biomass.

8.8.2. Rainwater Harvesting to Improve Farm Ecosystem Services

Existing research and farm-level and regional development programmes aimed at improvement of the rainfed systems have shown that proper development and use of the water harvesting system is the first entrypoint for success for most of these initiatives. The benefits associated with all additional activities concerned with improved soil and land management, such as crop and pest and disease management; investments in fertilisers, machinery and other agricultural investments; and development and access to markets, accrue to the field or the region which has a guaranteed access to the water resource.

Rainwater harvesting and its application to achieving higher crop yields encourages farmers to add value and diversify their enterprises. In parts of Tanzania, rainwater harvesting has enabled farmers in semi-arid areas to exploit rainfed farming by growing a marketable crop. Farmers upgraded from sorghum and millet to rice or maize, with additional legume crops that exploit residual moisture in the field. Similarly, studies of the Rajsamadhiyala watershed in Gujarat, India revealed that public investment in rainwater harvesting enabled farmers to invest in wells, pump sets, drip and sprinkler irrigation systems and fertilisers and pest management.

In addition, farmers in the developed watershed villages in Andhra Pradesh, India allocated a greater area to vegetables and horticultural crops than did the farmers in the surrounding villages, which contributed to income stability and resilience. Farmers also improved livestock and moved towards keeping large dairy animals (buffaloes, cows) rather than small grazing animals (sheep, goats). In this regard, the World Bank notes that each 1% growth in agricultural yield brings an estimated 0.5-0.7% reduction in rural poverty. Thus water harvesting improves agricultural productivity with more value added outputs and boosts rural employment, both on and off the farm.

When on-farm productivity increases, thereby improving rural incomes and human well-being, other ecosystem services can improve too. This has been especially obvious where in situ rainwater harvesting has been implemented to reduce soil erosion. Increased farm productivity can reduce pressures on forestry and grazing, and thus increase habitats and biodiversity.

Introducing rainwater harvesting to improve soil ecosystem productivity in rainfed agriculture promises large social, economic, and environmental paybacks, particularly in poverty reduction and economic development. Rainwater harvesting presents a low-cost approach for mediating dry spell impacts in rainfed agriculture. Remarkable successes have in fact been witnessed in poverty- stricken and drought prone areas in India and Africa. In Sub-Saharan Africa, the future of over 90% rainfed farmers depends heavily on improved water security. In South Asia, about 70% of agriculture is rainfed and some good work has been done in the design and successful demonstration of a range of water harvesting structures, for both drinking water supply and irrigation.

In several other countries in the Middle East, Latin America and South East Asia, rainwater harvesting is a traditional practice in certain regions, but the transferability of these models and practices has so far been limited. One of the main problems is that the local institutions needed often are inconsistent with the predominant governmental structures and institutional arrangements prevailing in these countries.

Investment in rainwater harvesting is important in meeting not only the Millennium Development Goals on reducing hunger but also on reducing poverty and ensuring environmental sustainability. A review of 311 case studies on watershed programmes in India, with rainwater harvesting and rainwater management as important components, found that the mean cost-benefit ratio of such watershed programmes was relatively high at 1:2.14. Rainwater harvesting created new/additional sources of water and helped in the provision and regulation of the water supply systems.

Poor management of rainwater in rainfed systems generates excessive runoff and floods, causing soil erosion and poor yields. Investment that maximises rainwater harvesting, both in situ and *ex situ,* helps to minimise land degradation, while increasing the water available for productive use. Investment is now needed in water resource management, in smallholder rainfed farming systems, which adds additional freshwater and improved soil storage capacity and productivity.

A small investment for supplemental irrigation, in combination with improved agronomic management, can more than double water productivity and yields in small-scale rainfed agriculture. This will release pressure on surface water or ground water for irrigation. It will make more water available to sustain aquatic ecosystems, without compromising agricultural productivity. Increased rainwater infiltration can also artificially recharge the depleted groundwater aquifers in hard rock regions and in areas of intensive groundwater use. Governments in India and Pakistan have developed elaborate master plans for the artificial recharge of the aquifers through recharge wells, recharge shafts and recharge ponds.

Environmental and social concerns need to be given due consideration when implementing rainwater harvesting projects. In basins with limited surplus supplies, rainwater harvesting in the upstream areas may have a damaging impact downstream and can cause serious community conflict. Also, when runoff is generated from a large area and concentrated in small storage structures, there is a potential danger of water quality degradation, through introduction of agro-chemicals and other impurities. Special investigations on water quality must be undertaken before using the harvested water for recharge of underground aquifers.

8.9. Forests Ecosystems as Water Harvest Interventions

Net deforestation has lost some pace during the last decades, but is still severe in a global context. Notably, the range and nature of deforestation is very variable in different regions and countries. In many cases, with intensifying cultivation and conversion to pasture or permanent low-input agriculture, the result is not only loss of biodiversity and its related ecosystem services, but landscapes are at risk of erosion, water pollution, flooding and decreasing soil productivity. These land use and land quality developments are very undesirable from the perspective of meeting the needs for increased biomass production for food and energy, as well as for ensuring a supply of clean water.

On the other hand, much less attention is given in the media to the simultaneous processes of increasing forested areas in some regions and the increasing use of planted trees for various purposes. Planted forests have historically contributed to development in many countries in temperate regions, and have the potential to improve the livelihoods of millions of people in other regions.

Today, planted forests comprise 6.9% of the world's total forest area of which more than half is located in the South. In 2050, FAO predict that 75% of global wood consumption will come from planted forests and that this expansion will be global. Recent expectations of forests as bio-energy reserves may dramatically raise the demands for new planted forests. 75% of planted forests are intended for industrial production. Forests owned by smallholders increased more than 3 times during 1990-2005 and now represent over 30% of all planted forests.

Outside these figures, trees planted outside forests and on homesteads are increasing steadily. Apart from FAO definitions of planted forests, this group represents a continuum of use of trees for a variety of purposes in small woodlots, agroforestry and homesteads. Most of the small holder increase is in Asia. In Africa there is a significant increase in timber plantations. In the near and mid-term future, these plantations will continue to expand, driven primarily by the growing demand from China and India. In recent years, both countries have invested heavily in timber plantation holdings, both nationally and overseas.

Agroforestry, or systems of intercropping permanent and annual crops, has gained a positive aura and developed strongly to improve traditional cultivation systems in a broad variety of environments. The relative success of biomass production in planted forestry has in many cases been overshadowed by negative ecosystems impacts and social-institutional issues. Ecosystem services affected include shifts of water use within the landscape and losses of biodiversity when converting from natural forest. When established without consideration of local stakeholders exclusion from previous livelihoods, it has sometimes caused longstanding conflicts. However, as the natural forest cover continues to degrade and decrease, there is an increasing need for planted forests.

Forests, trees and bushes form specific part of landuse systems. Considering the water balance, tress normally uses more water per area than an annual cereal crop in the very same location. Thus, the 'old paradigm' of forests as 'water towers' or as 'water protectors' is rarely valid in the landscape. However, the provisional ecosystem services capacity of a woodlot, apart from water, can outweigh those of the same area being cultivated. In general, the total biomass gain is higher and biodiversity is improved, provisioning a range of produce which can be harvested, often more reliably than annual crop systems. Forests also provide wood and energy. From a regulatory perspective, trees and forests play a significant role in affecting soil infiltration capacity and reducing erosion. They enhance soil quality through litter fall and extensive root systems, and have been shown to act as water purifiers.

Trees and forests in many cultures often fall under special local management systems, to ensure their sustainable maintenance. Often, trees and forests are associated with high aesthetic and spiritual values. Thus, from a comprehensive livelihood perspective, forests and trees in the landscape offers multiple ecosystem services for the water consumed. Many of these products are essential in times of crop failure, when forest products can provide food and income in times of crisis.

It is now an empirically and theoretically well established general scientific paradigm that forests use more water than lower vegetation and annual crops in rainfed agriculture. Consequently, empirical evidence is strong that cutting forests results in increased stream flows. Typically, when forest cover is regenerated, more rainfall tends to be partitioned through soil infiltration and to green water, reducing its availability as blue water downstream.

As a special case in semi-arid areas, old growth forests may work as "sponges" to better retain or recharge groundwater and to maintain dry season stream flow. This has long been an item for scientific and policy debate. Forests have been shown to maintain a high soil infiltration capacity by superior litter fall and soil protection. Increasing surface runoff after deforestation increases surface run-off and possible soil deterioration, leading to more "blue water," but water that is often polluted by soil erosion.

The higher surface runoff during rainfall events at a deforested location means that less water is contributed to long term groundwater recharge on site during the wet season. Depending on the location, shallow groundwater is often linked to lower lying stream flows, regulating the river base flow during dry seasons.

Decreasing shallow groundwater recharge through deforestation may thus deplete surface water sources in times of high demand. The reduction of stream flow after deforestation has often been observed by rural people, but only a few studies have reported the expected long term decline in dry season flows. Thus, we have some evidence that a "sponge effect" can be lost by deforestation and subsequent soil degradation, but the conclusion can hardly be made general for all semi-arid forest ecosystems.

Based upon evidence on how tree litterfall and soil protection can improve soil quality and reduce surface runoff and erosion, the restoration of a "forest sponge effect" has generally been taken for granted. This has been the paradigm behind numerous forest/tree planting projects and one of several drivers for adoption of agroforestry. However, in this case, there are many local witnesses to the fact that new forests often make wells and streams even drier than after deforestation.

As for scientific studies in this case, long term studies are scarce. In contrast to the "lost sponge effect" paradigm, the few studies conducted in semi-arid environments all confirm that new forests use more green water than they contribute to blue water in terms of groundwater recharge. This effect of "not enough ground water recharge" is manifested in these studies as generally declining stream flows following (re)forestation.

High water use by new forests reflects higher production. The new forests established are most often planted exotic species like eucalyptus and pines. They are chosen for their high productivity. Many of the species used are pioneer species in their respective original ecosystems, and increasingly they are genetically improved for fast wood production, but not necessarily to be water efficient. Furthermore, these new forests are monocultures of vigorously growing young trees in contrast to old growth forest, which are mixes of species, old trees, young trees and treeless gaps.

Deep rooted eucalypts are often given as an "example" of the highly water consuming exotics, but a range of other tree species may show similar relative increases in water use, compared to the natural forest in a given site. In South Africa, the water consumption of trees is recognised in water management. To establish a wood lot or plantation requires special permission from local forest and water authorities, and is associated with specific fees and costs as it will decrease water available for other uses in landscape. One reason for increased water consumption in afforested areas is the use of exotic, more water consuming species, compared to the native vegetation.

We conclude that forest water use is often a significant factor in landscape water flows, including surface, subsurface and downstream. But the specific impact on the water resources of deforestation and afforestation is governed not only by site specific soil and topographic conditions, but also by whether species are native or exotic; whether trees are in large homogenous plantations or in a landscape and stand structure mosaic. The water use and partitioning of a forest stand is also relative to the site's natural or alternative landuses and water balance flows. Due to the complexity of forests and their impacts on the local water balance, few comprehensive case studies exist for each climatic, vegetative and hydrological response, especially for semi-arid tropical regions with previously forested, now degraded, soils. In contrast the few studies available are from southern Africa and India where former non-forested grasslands and savannas have been afforested. Thus, the lack of data and empirical evidence is seriously challenging our ability to assess potential water balance impacts by deforestation or afforestation in specific landscape contexts.

8.9.1. Synergies and Trade offs in Miombo Woodlands

Miombo woodland is a significant biome covering about 10 % of the African landmass, approximately 2.5-4 million km^2 depending on definition. It supports the livelihood of 100 million people in the area or outside, relying on products from this distinct and unique biome. Major provisional ecosystem services essential for livelihoods are charcoal for rural and urban energy, water for downstream needs, wood, meat from grazing and hunting, fruit, tourism, and habitat provision, etc. In addition, the woodland affects several regulatory services, such as landscape water flows, soil erosion control and regeneration of soil health in the smallholder systems. Throughout its physically varied region, miombo woodlands overlap with deciduous forests and open savannahs.

The climate is semi-arid with one wet season, but annual rainfall ranges as much as 550-1200 mm and dry season lasts between 3 and 7 months. The miombo woodlands also coincide with some of the poorest sub-Saharan African countries, with relatively low rates of achievement of many Millennium Development Goals relating to water supply and sanitation. The high prevalence of HIV/AIDS among other diseases is a big challenge to the people living in the area.

Deforestation is an old and ongoing process in the miombo region, but large areas are still covered by miombo in various states. Long term human impacts are often profound on forest structure and species composition in many areas. Forest management and tree planting mostly has been focused on exotic species in plantations and woodlots, even if, more recently, there are increasing numbers of interesting examples of natural forest management in Zimbabwe and elsewhere. Tanzania and Zimbabwe are the central concerned miombo countries that have the most forest plantations. Total areas are still moderate and about half of them are industrial.

Looking ahead, with increasing demands for energy, industrial wood and carbon credits, there is a growing interest in plantation forestry in the relatively sparsely populated miombo region, not least in Tanzania. The miombo landscape provides a very varied structure and net primary productivity of the continuum ranges from degraded miombo to well-managed miombo to even-aged forest plantations. This has large impact on water management, both through water use by the trees as well as by the impact on soils and potential groundwater recharge. Any major change in the miombo woodlands needs serious consideration: can the 'sponge effect' be lost? And what implications does that have on provisioning and regulating ecosystem services supporting vulnerable livelihoods in the area?

An example of the altered water balance due to the planting of exotic species is the decreased lower annual and dry season stream flows in areas populated by *Eucalyptus saligna* compared to nearby grasslands in Sao Hill, Tanzania; the stream flow reduction by *Pinus patula* was much less. Thus, planting eucalyptus could potentially mean a loss of biodiversity, and reduced dry season flows. In another study, using long term data, Kashaigili *et al.* show major decreases in dry season flows (60-70%) between 1958 and 2004, downstream of the Usangu wetlands in Tanzania. In the upstream areas, woodlands have decreased strongly due to expansion of cultivated lands and bare land. In this case it may be tempting to hypothesise on the "lost sponge effect", but Kashaigili *et al.* used modelling to show that the major reason for declining dry season flows was due to increase of irrigated agriculture upstream from the wetlands.

8.9.2. Healthy Soil Sytem for Catching Rainfall

Management of organic material in soils is crucial for a healthy soil system. Soil organic matter influences soil physical characteristics and availability of plant nutrients. Increased soil organic matter increases soil water storage capacity, and water infiltration capacity.

Harvesting, grazing and fire lead to degradation by reducing litterfall; i.e., contributing to reduced organic matter content, and oxidation. In miombo, already low topsoil organic contents are typically reduced by up to 50 % by agriculture. Soil organic matter also determines top soil physical properties. The soil structure determines to a large extent the partitioning between surface runoff, erosion and soil water infiltration.

In various land uses in Zambia, the structural stability of the soil was shown to be positively related to soil organic carbon. Soil crusting is a common reason for reduced soil water infiltration in semi-arid areas. Perrolf and Sandström concluded that vegetation cover was the other major determinant apart from soil texture in Tanzania and Botswana. Similarly, Casenave and Valentin, using data from 87 sites in semi-arid West Africa, found intensity of surface sealing, vegetative cover and soil faunal activity to be determinants of soil water infiltration. Organic matter in the soil is highly dependant on vegetation species composition. Research suggests that miombo may not always be superior to exotic species, but the condition of the miombo stand does affect the litter, the soil organic matter and subsequently the infiltration of rainfall.

In the miombo region, like other semi-arid areas, a higher intensity of land use in trees is already leading to environmental degradation. Despite inadequate scientific clarity in regard to the biophysical processes and lack of empirical data, resources have to be managed. Multi-species plantations in general are shown by meta-analysis to be more productive than mono-specific plantations. At the same time these more complex forest stand types might have a more favourable impact on infiltration and a more moderate water demand compared to most even-aged monocultures. In addition, experiences of development of smallholder involvement in forest establishment and management from Asia might be fruitful to apply in the miombo region.

8.9.3. Parklands—Role of Trees in Generating Soil and Water Gains

Sudano-Sahelian parklands stem from dry deciduous forests with some relation to miombo. These parklands have had strong human influence on the structure of the vegetation for a long time. While small-scale shifting cultivation is dominant in miombo, the parkland of West Africa is dominantly under permanent traditional agroforestry with dense wooded savannah. There is an abundance of preferred indigenous tree species. The fallow periods are continuously being shortened because of pressure for land. However, farmers retain trees in farmlands for their own livelihood purposes.

Certain species such as *Vitellaria* provide valuable butter from the kernels of the tree nuts. This is used for local consumption and provides an important source of income for rural women. Products from several tree species are also used in traditional medicine and produce edible fruits. However, due to the intensification of agriculture in the region, the parklands are decreasing and in many cases the tree cover is diminishing. Several studies have shown' that trees add soil organic matter through litter fall as well as promoting biological activity in the soil and thereby improve the physical properties of the soil. These benefits are similar to the effects of applying compost manure to the fields.

Little is known from research about the effects of trees on the water management of the parklands, although vegetative cover in the region is considered to be important to ensure maximum rainwater infiltration into the soil profile. In addition, Bayala *et al.* concluded from a study carried out in the parklands of Burkina Faso that an application of leaf litter mulch from *Parkia biglobosa* and *Vitellaria paradoxa* prunings improved soil organic matter content as well as water infiltration. More recent Bayala *et al.* have shown some parkland trees to hydraulically redistribute water during the dry season. This means that the trees at night transport water from deeper soil layers to the top soil. This is beneficial for both the trees and other plants during hot dry season days. However, as for miombo, there has not been clear scientific verification of the effect of the trees on water budgets. The effect on rainfall by re-introducing trees and their management in the parklands also has not been clearly synthesized and interpreted.

On a field scale, *in situ* rainwater harvesting can enhance re-establishment of trees on the landscape. The success of re-greening of the Central Plateau, Burkina Faso is an example where land reclamation through *in situ* water harvesting has led to increased numbers of trees on former crop-land. In this case, the severe droughts generated a positive response in terms of activating communities and mobilising resources to address multiple challenges including poverty, low crop yields and severe land degradation.

Farmers, NGOs, local government and external funders together enabled the adoption of *in situ* rainwater harvesting for growing crops and trees. A few key technologies were particularly interesting such as the *zai* pitting, and the construction of stone bunds and gully control structures. The results of *in situ* rainwater harvesting showed that average crop yields doubled over 20 years, producing more forage leading to increased livestock numbers and establishment of more trees.

In addition, species diversity in fauna was regenerated, and a noticeable rise occurred in the groundwater table. Although soil fertility has improved through better soil and water management, there is more potential for improvement. According to the local communities, food security has improved to meet the demand of the population, that has increased by 25% between 1984 and

1996. An important effect of improved yields is that no further crop land expansion has occurred since mid 1980s despite the population growth and improved livelihoods.

Removal of trees and lack of regeneration in the parkland is often driven by the introduction of mechanised farming in a cotton and maize rotation system. This is for increased food production and cash incomes for local communities. However in the long term, this production system may lead to a decline in soil organic matter, fertility, high erosion risk and soil degradation.

Maintaining soil organic matter is important for carbon sequestration and better moisture retention for improved crop harvest. This is instrumental for climate change adaptation where scenarios indicate drier climates in years to come. In contrast to increasing demands for higher biomass production and increased crop yields, and in view of the lack of reliable data on trees and their benefits, it is not easy to motivate people on economic and long term benefits of trees in parklands.

References

Agarwal A.,and S. Narain. *Dying wisdom: Rise, fall and potential of India's traditional water harvesting systems* 4th edition. Eds., State of Indias Environment, a citizens' report 4, Centre for Science and Environment, New Delhi, 2005

Barron, J., Noel, S. et al. *Agricultural water management in smallholder farming systems: the value of soft components in meso-scale interventions.* SEI Project Report, Stockholm Environment Institute, Stockholm. 2008

GEO4. Global *Environmental Outlook 4: Environment for development.* United Nations Environment Programme, Nairobi/ Progress Press, Malta. 2007

Millennium Ecosystems Assessment (MA). *Ecosystems and human well-being: synthesis.* Island Press, Washington D.C. 2005.

Kumar, M.D., Patel A., Ravindranath R., Singh O.P. "Chasing a Mirage: Water Harvesting and Artificial Recharge in Naturally Water-ScarceRegions", in *Economical and Political Weekly,* August 30, 2008.

9 Rainwater Harvesting for Human Development

Water is essential to all life forms on earth—human, animal and vegetation. It is therefore important that adequate supplies of water be developed to sustain such life. Development of water supplies should, however, be undertaken in such a way as to preserve the hydrological balance and the biological functions of our ecosystems. Consequently, the human endeavour in the development of water sources must be within the capacity of nature to replenish and to sustain. If this is not done, costly mistakes can occur with serious consequences. The application of innovative technologies and the improvement of indigenous ones should therefore include management of the water sources to ensure sustainability and to safeguard the sources against pollution. As land pressure rises, cities are growing vertical and in countryside more forest areas are encroached and being used for agriculture.

In India the small farmers depend on Monsoon where rainfall is from June to October and much of the precious water is soon lost as surface runoff. While irrigation may be the most obvious response to drought, it has proved costly and can only benefit a fortunate few. There is now increasing interest in the low cost alternative-generally referred to as Rain Water Harvesting (RWH). Water harvesting is the activity of direct collection of rainwater, which can be stored for direct use or can be recharged into the groundwater. Water harvesting is the collection of runoff for productive purposes. Rain is the first form of water that we know in the hydrological cycle, hence is a primary source of water for us.

Rivers, lakes and groundwater are all secondary sources of water. In present times, we depend entirely on such secondary sources of water. In the process, it is forgotten that rain is the ultimate source that feeds all these secondary sources and remain ignorant of its value. Water harvesting is to understand the value of rain, and to makc optimum use of rainwater at the place where it falls. Availability of water is critical for ecosystem health and productivity, ensuring supply of a range of products and services, to benefit human well-being. With growing multiple demands of water, the ecosystems supporting and regulating the structure and function of natural ecosystems may be eroding. There is an urgent need to find opportunities to enable development and promote human well-being without undermining ecosystem health.

Rainwater harvesting locally collects and stores rainfall through different technologies, for future use to meet the demands of human consumption or human activities. The art of rainwater harvesting has been practised since the first human settlements. It has been a key entry point in local water management ever since, buffering supplies of rainfall to service the human demand of freshwater. As it involves the alteration of natural landscape water flows, it requires water managers to carefully consider the tradeoffs; however, it can create multiple benefits, offering synergies between different demands and users at a specific location. To many water managers, rainwater harvesting is a technique to collect drinking water from rooftops, or to collect irrigation water in rural water tanks. However, rainwater harvesting has much wider perspectives, in particular if it is considered in relation to its role in supporting ecosystem goods and services.

9.1. Rainwater Harvesting for Rural Well-being

At the 2000 UN Millennium Summit, world leaders from rich and poor countries alike committed themselves to eight time-bound goals as a blue print to accelerate development. The resultant plan is set forth in the Millennium Development Goals (MDGs). Goal 7 addressed the environment and water. Its targets include the goal to "halve, by 2015, the proportion of the population without sustainable access to safe drinking water and basic sanitation". In relation to water, this implies provision of safe water for drinking as well as for hygiene. And because these amounts are relatively small (compared with e.g., agriculture) there are large potentials to exploit rainwater harvesting in this context of human well-being.

Halfway through to the 2015 targets, globally, the target related to drinking water is expected to be met, but not the sanitation target. Nevertheless, these global figures mask a critical situation in some regions. For example, sub-Saharan Africa and Oceania will not meet these targets at the present rate of implementation. With 2.4 billion persons without access to sanitation in 2004, the target will also be missed, particularly in South Asia, East Asia and sub-Saharan Africa. To meet the target requires doubling the efforts of the last 15 years for sanitation, and increasing those made for drinking water by a third.

Though only one target addresses water, it plays a critical role in meeting the goals, particularly those concerning hunger, poverty, health and biodiversity. To meet the MDGs often assumes a reliable source of good quality water. As an example, the MDG water target focuses on infrastructure, policies, awareness creation, etc. with little attention to the sustainable management of the water sources.

Available freshwater continues to decline due to over-abstraction, pollution and reduced precipitation, resulting in a decrease in runoff. An estimated 1.8 billion will live in water scarce areas and two- thirds in water stressed areas by 2050. Climate change will worsen the situation. Ecosystems play an important part in water availability and vice versa, but the link between ecosystems and water availability is complex and not fully understood.

Freshwater ecosystems link directly and indirectly with human well-being, especially the well-being of poor communities and households. There is a close interdependency between freshwater ecosystems and human well-being. Though it is not easy to put an economic value on ecosystem services, some attempts have been made. The Millennium Ecosystem Assessment found that 60% of the ecosystems assessed were in global decline, particularly the aquatic ones, with detrimental effects to human well-being. A significant cause of ecosystem degradation is overabstraction for water supply. Rainwater harvesting can be used to improve ecosystem function, particularly the water supply aspect, and regulation (controlling flood and erosion). Globally, there is a significant untapped potential in rainwater which can be harvested to improve ecosystem services and human well-being.

9.1.1. Rainwater Harvesting and Rural Water Supply

Most people living in rural areas depend on development which is based on ecosystem services. Typical rural ecosystem services that support human livelihoods include water supply, agriculture including livestock management, fisheries, and forest and tree products (timber, honey, fruit, vegetables, fibres, fuel etc.). In many parts of Africa, wild fruit, firewood and charcoal are major sources of income, especially in times of crop failure. Ecosystems are also necessary for water purification, erosion regulation, waste treatment and disease regulation. Support services include soil formation (providing good soils for agriculture and vegetation).

Degradation of these ecosystem services is threatening the achievement of the MDGs. About one sixth of the world population – a total of 1.1 billion people – remains without access to improved drinking water, and 84% of these live in rural areas. In addition, 2 of the 2.6 billion people without access to basic sanitation live in rural areas. The figure differs according to regions. For example, in 2004, in sub-Saharan African rural areas, the number of people who were not provided with improved drinking water was five times higher and those without access to sanitation were three times higher than those living in urban areas. Poverty is also much higher in the rural areas.

In rural communities, water is required for drinking and agricultural purposes. Rainwater harvesting is highly decentralised and enables individuals and communities to manage their own water for these purposes. This is particularly suitable in rural areas with a dispersed population and where a reticulated water supply is not feasible or extremely costly for investment.

The low cost of the rainwater harvesting technologies can be a more attractive investment option in rural areas. In addition to water for drinking and sanitation, fishing, animal husbandry and agriculture are the major activities in rural areas which all depend on a reliable water supply to be productive.

Rainwater harvesting can also help meet the demands for water for these purposes. There are numerous cases where rainwater harvesting is used to improve livelihoods by providing water for domestic purposes; for subsistence and income generation activities such as gardening, and livestock rearing; for environmental purposes, through recharging groundwater and establishing woodlots to reduce deforestation. In essence, it can supply water to accelerate social and economic development, to alleviate poverty and generate income for rural farmers by enhancing the crop yield, modifying the method of production, as well as to promoting environmental conservation.

9.1.1.1. Role of Agro-pastoralists

In Kaijado and Lare, in the semi-arid savannah of Kenya, rainwater harvesting provides water for drinking, sanitation, and enhancing the productivity of the agro-eco systems. The technologies introduced consisted of roofwater harvesting for domestic purposes (drinking and sanitation), runoff collection in ponds for small gardens, trenches for groundwater recharge and afforestation. For sustainability, the project included a micro-finance component, where the community was trained to manage credits before borrowing money from commercial institutions.

The project has enhanced the ecosystem functioning by recharging groundwater, increasing the volume of water stored, and reducing soil erosion through the family woodlots that reduced runoff-related erosion. Once the planted trees have matured, the women will use them for fuel, contributing to the reduction of deforestation, which is a major problem in the area. Family livelihoods improved from selling vegetables and income generation activities such bee-keeping and crafts.

The community can now borrow money from commercial micro-finance companies which they use for productive activities. Since the establishment of the micro-finance component, all of the members have been paid on time, and there have not been any arrears. Providing water to schools enabled girls to attend during their period of menstruation, thereby increasing their attendance.

9.1.1.2. Adapting Water Supply through Rainwater Harvesting

The semi-arid region of Brazil (SAB), in the northeastern part of the country, has a rainfall which can range from below 185 mm to 974 mm between one year and the next. It is concentrated within a few weeks of the year and is associated with a high evaporation rate of 3,000 mm a year. In 2005, the Ministry of National Integration calculated its drought risk between 1970 and 1990 as being above 60%. Climate change forecasts indicate that the drier parts of the SAB will become even drier, despite there being a small increase in precipitation.

To adapt to the current rainfall variability, more water storage is needed in rural areas. Rainwater harvesting is one way to adapt to current and future rainfall variability. The Programme for 1 Million Cisterns (P1MC) was initiated to supply safe drinking water for 1 million rural households (five million people). With funding from the government and the private sector, more than 230,000 cisterns were constructed as of August 2008 with some municipalities constructing their own.

Evaluation of the programme found that the health of the population improved through better drinking water quality and time saved for women, who no longer need to fetch water over long distances to their homes. From an ecosystem service perspective, rainwater harvesting and storage has impacted farm productivity in numerous ways. Using water for the irrigation of higher value crops, such as in kitchen gardens, especially off-season has been beneficial for household food supplies and incomes. Rainwater harvesting has also resulted in an increased number of goats per household, partly as more fodder is available.

On the land, observed changes include: reduced erosion through the practice of conservation tillage and construction of soil bunds, reduced flooding downstream, and increased species diversity due to infiltration banks and sub-surface storage dams. So far, the effects of rainwater harvesting have not affected water supplies downstream.

Rainwater harvesting has been accepted by the rural community in the SAB who have learned to live in harmony with nature in a semi-arid climate, and are ready to fight for it, as well as for all the other aspects which might improve their livelihoods. They understand that water must be managed in an integrated way, taking into consideration the source (rain, surface water, soil and ground water), and water uses (for environment, domestic, agricultural and emergency purposes).

9.1.1.3. Small Rainwater Storages

Gansu Province is one of the driest, most mountainous and poorest regions in China. It has an annual precipitation of 330 mm while potential evaporation is as high as 1,500-2,000 mm. Rain is the only water available and reticulated water systems are not feasible because of the terrain and the sparse population. This is an area lacking in three essentials: water, food and fuel. This causes insecurity in both human livelihoods and the environment. From 1988 to 1992, research was conducted to find the most suitable rainwater harvesting interventions to promote in the area. By the end of 1994, 22,800 updated water cellars with 2.4 million m^2 of new catchment area (tiled roof and concrete lined courtyard) had been built. A total of 28,000 families (141,000 people), 43,000 large livestock and 139,000 small animals got enough water to drink.

Rainwater harvesting also played a significant role in promoting ecological and environmental conservation. The "Land Conversion" Programme for the north and northwest China, along with development of rainwater harvesting, increased the area of orchards. In the period 1996 to 2000, 73,300 ha of land were irrigated with water supplied through the rainwater harvesting, with the irrigated areas used for the planting of trees in Longnan Prefecture.

In Yongjing County, 273 ha of trees were planted, using the rainwater harvesting system. The productivity of farm land has increased, with the introduction of the rainwater harvesting systems. Yields have improved by 20-40% in the fields, kitchen gardens and holdings of pigs and sheep. Species diversity has increased, since rainwater harvesting enables greater diversity of crops and cropping patterns. As incomes rise, people no longer need to degrade their landscape to support livelihoods.

One key improvement in the environment introduced through rainwater harvesting is reduced soil erosion, which maintains better soil quality on site, and reduces siltation in waterways and dams downstream. Reduced incidences of flooding downstream indicate another positive effect of the rainwater harvesting interventions upstream. The rainwater harvesting approach that has been adopted since the late 1980s has brought about tremendous changes in the rural parts of the dry, mountainous areas of China. Experiences with rainwater harvesting during the past 20 years show that rainwater harvesting is a strategic and invaluable measure for achieving integrated development in the rural areas. Statistics show that, by adopting rainwater harvesting techniques, 15 million people have solved their drinking water problems and 2.6 million ha of land have been irrigated. Other intermediate techniques such as rudimentary greenhouses, solar heating, and the indigenously innovated underground tanks have also been adopted.

Use of rainwater harvesting systems for domestic water supply, agriculture and drought mitigation has spread to the semi-humid and humid areas of China that suffer from drought, such as Southwest China, the coastal towns of Southeast China, the islands and Guangxi Autonomous Region.

9.1.2. Cost-effectiveness of Rainwater Harvesting

The world's urban population increased from about 200 million (15% of world population) in 1900 to 2.9 billion (50% of world population) in 2000, and the number of cities with populations in excess of 1 million increased from 17 in 1900 to 388 in 2000. As people increasingly live in cities, and as cities act as both human ecosystem habitats and drivers of ecosystem change, it will become increasingly important to foster urban systems that contribute to human wellbeing and reduce ecosystem service burdens at every level.

Severe environmental health problems occur within urban settlements, resulting from inadequate access to ecosystem services, such as clean water. Many ecosystems in and around urban areas are more biodiverse than are rural monocultures, and they can also provide food, water services, comfort, social amenities, cultural values, and so on, particularly if they are well managed. Moreover, urban areas currently only account for about 2.8% of the total land area of Earth, despite containing about half the world's population.

9.2. Impacts of Urban Hydrology on Ecosystems

Cities require large amounts of water to sustain themselves, owing to the sheer size of population and density of housing. Up to now the most typical method of meeting this demand has been to build a large dam or withdraw water from groundwater sources and pipe it in as needed. This can be ecologically disruptive, as well as costly, causing water stress in the river downstream, and changing the biodiversity of the region. In addition, groundwater levels may decline in urban areas as a result of increased pumping as well as extensive areas of impermeable surfaces, with hardly any natural infiltration.

In urban areas rainwater is disposed of most commonly as storm flows, in underground pipes, and as quickly as possible. Concentrated storm flows can alter surface water flow patterns, affect flora and fauna, and potentially increase the risk of flood damage downstream. The lack of surface water in urban environments can cause the dry micro-climate to increase in cities, the so called 'heat island' effect. This leads to a greater need for cooling systems.

If climate change brings higher peaks and intensities in the volume of precipitation, current stormwater drainages will be too small to cope and more incidences of flooding may occur. The problem of financing the water supply and wastewater infrastructure will increase as a result of demographic changes and urban-rural settlement shifts. However, we have to adapt our technologies of urban hydrology to this change in demography and climate. This is our chance to modernise these very important infrastructure elements for the benefit of the economy and society, by making them decentralised, and thus more affordable. Cities have access to rainfall, which can be used to supplement water abstractions from surface or groundwater sources and help meet demand.

9.2.1. Water Quality and Health

Surface water is often contaminated through the release of

industrial and domestic effluents directly into lakes and rivers, and from pesticide and agro-chemical run-off from fields. In theory, rainwater is the safest of all water sources. Although rainwater can become contaminated through the absorption of atmospheric pollutants, it is usually clean as it hits the earth, unless there is atmospheric pollution from industry.

The challenge with rainwater is to keep the collection surfaces (roof tops) and the storage facilities free from contamination and free from mosquito breeding. With the adequate operation and maintenance of the collection areas, filter and tank systems, good quality water may be obtained by collecting rainwater from rooftops. While high-quality source water may require little or no treatment, it is still recommended that any water used for drinking be disinfected to ensure microbiological safety. According to the Water, Sanitation and Hygiene programme of WHO, maximum health benefits are achieved if water interventions are accompanied by sanitation and hygiene promotion.

9.2.2. Synergy Advantages

Urbanisation puts the surrounding water resources under pressure, challenging ecosystem services in two principal ways. Firstly, the concentrated urban population demands adequate water for consumption and sanitation needs, which requires stable and large supplies of water, often through the use of surface water/dams or groundwater. These extractions can threaten other landscape habitats and functions, reducing the ecosystem's capacity to supply things such as water downstream, habitat for biodiversity, and livelihood support. Secondly, the reduced infiltration of urban landscapes alters the flow downstream, and can increase the incidence of flooding. Rainwater harvesting in urban areas can address both these negative effects. Rainwater harvesting tanks contribute to the redistribution of flows over longer temporal scales, thus reducing the incidence of flooding downstream. The additional effect, is that the collected rainwater is used, which means that demand on other water sources can be reduced. Many synergies between water storage and additional non-accounted positive effects have been found for rainwater harvesting projects in urban areas:

- evapotranspiration from planted roofs, retention swales and ponds, cooling down the urban heat island effect in cities, thus improving human well-being and saving energy for cooling in hot weather periods (reducing the CO_2 footprint, mitigating climate change)
- energy savings due to reduced pumping
- aesthetic values of architectural and landscape water features, created by providing rainwater retention on site
- biodiversity values of planted rooftops, retention swales and ponds, provided by rainwater retention on site, improving ecosystems even in urban areas
- infiltration from retention swales and ponds, recharging groundwater aquifers
- daylight reflection from retention ponds designed close to buildings, lighting up building interiors, thus helping to save energy (reducing the CO_2 footprint, mitigating climate change)
- greater awareness of the ecosystem, when citizens run their own rainwater harvesting system, thus saving not only water, but also other natural resources, e.g. energy. Also usable for education in schools and universities (reducing the CO_2 footprint, mitigating climate change)
- higher concentration of sewage water in mixed sewers, improving the functioning of centralised sewage treatment plants resulting in cleaner outflows to rivers (reducing the impact on ecosystems) and less pumping energy expenditure in the plants, thus reducing costs for the community (enhancing human well-being in socio-financial terms) and mitigating climate change (reducing the CO_2 footprint).

9.2.3. Revival of Rainwater Tanks

Australia is a country that can look back on a long tradition of rainwater utilisation. In recent years, Australia has been facing a water crisis. Increasing population, cheap water, and a failure to add new supplies, exacerbated by the effects of climate change, have brought home a stark reality: some cities have been running out of water. The severe water restrictions placing harsh limitations on the watering gardens and washing of cars along with a strong personal sense of wanting to do something about the water crisis have led to a huge revival in domestic rainwater tanks. Spurred on by generous rebate schemes, Australians just love them. Rainwater tanks have become the latest "must have" item. This has now spilled over into commercial rainwater harvesting. For new buildings, thanks to the Green Star rating scheme, it is almost a pre-requisite to install a rainwater harvesting system.

Approximately 30% of Australia's urban water consumption is non-residential. A quarter of that could be reduced through water efficiency measures. Of this demand, some 8% could readily be supplied by commercially viable rainwater harvesting schemes. Such projects can capture rainwater from 1,000 to 10,000 m^2 and more. If schemes collecting storm water (i.e. rainwater including ground surface runoff) were included, collection areas of 50,000 m^2 and beyond could be achieved.

Large rainwater harvesting schemes are of interest to hospitals, works depots, shopping centres, tertiary institutions, military bases, prisons, sports facilities and parks and gardens. The goal is to build an integrated water supply network where the large dam supply systems work hand in hand with thousands of mini-dam supplies, in the form of both residential and commercial rainwater tanks installed throughout the municipal area. In Australia, it would also match the area where the greatest water demands occur with the regions enjoying the greatest rainfalls.

9.2.4. Decentralised Water Supply

The complete dependence of city water supply and drainage on a centralised system in the age of steady climate change and increasing urbanisation is precarious. There are also the issues of an ageing infrastructure and increasing energy costs. These natural and man-made risks can be reduced through the addition of a decentralised water management system.

Star City is a major real estate development project of more than 1,300 apartment units in Gwangjin-gu in the eastern section of Seoul. The basic design idea of the Rainwater Research Center (RRC) at Seoul National University and Professor Mooyoung Han was to collect up to the first 100 mm of rainwater falling on the complex and to use it for gardening and flushing public toilets. The entire fourth floor below the ground in Building B at Star City is used as a water storage area. Altogether it can store 3,000 m^3 of water, in three separate tanks, with a total floor area of 1,500 m^2. The capacity of each tank is 1,000 m^3.

The first two tanks are used to collect rainwater from the rooftop and the ground that mitigates the danger of a flood in the area during the monsoon season. Collected rainwater is used for the purpose of water conservation. A special feature is that most of the irrigated water in the garden is infiltrated into the ground and returns to the tank for multiple uses. The third tank is used to store tap water in the case of emergency. Fresh tap water is maintained by decanting half of the old water to the rainwater tank and refilling it on a regular basis. Based on the half year operation of the system, water conservation is expected to be approximately 40,000 m3 per year, which is about 67% of the annual amount of rainfall over the Star City complex. The risk of floods can be controlled pro-actively with the remote control system, by emptying or filling the tanks appropriately. The third novel concept applied in this project was the city government's incentive programme for the developer.

At a broader level, because the decentralised system harvests rainwater on site before it becomes dirty, it reduces the energy required—and therefore the carbon dioxide production and the long-term social cost—for water treatment and transportation. A cost comparison exercise on conventional and rainwater harvesting systems in Seoul City indicated that the energy required to treat and deliver a cubic metre of tap water is 0.2405 kWh, with most of this being energy for transmission. For grey water, treated on site, this would be 1.1177 kWh per cubic metre. According to Prof. Han's calculations, the same volume of rainwater, needing no treatment, can be delivered for a mere 0.0012 kWh, which is the pumping energy needed to raise the water from storage. The universal imperative is to provide water services with the lowest possible use of energy. Upon evaluation of the Star City project, the city government has already passed a city-wide ordinance to promote more rainwater harvesting system installations in development projects.

9.2.5. Taskforce for Environment

A broad range of medical and care services and 577 beds make the Bad Hersfeld Clinic the center of medical competence for eastern and central Hessen. With approximately 1,400 employees, it is one of the largest employers in the region. Both municipal and private investors are examining operating costs, especially the costs of energy and water. Rainwater utilisation in Germany often leads to a double benefit in terms of saving money; namely, the costs for potable water and fees for rainwater disposal.

In 1988, an "environmental task force" was established at the clinic. In the first phase of construction, in 1995, rainwater was already utilised for outdoor watering. In addition, a fountain and a pond were supplied with water from the cistern. Since 2001, 111 toilets have been connected to the rainwater system. The cooling of vacuum pumps used for sterilisation is especially effective. Rainwater, at a maximum temperature of 20°C, is circulated through the cistern in a closed system, where the waste heat is re-used. In 2007, with 384 m^3 of additional drinking water required during dry periods, the rainwater yield was then 2,180 m^3. To this we add the 4,000 m^3 of cooling water saved every year making a total of 6,180 m^3 of water conserved.

Since January 1, 2003, the clinic has also benefited from an amendment to the articles of the city of Bad Hersfeld. The new rate for rainwater per square meter of paved/sealed surface that runs off into the sewage system is 0.66 euros/m^2 throughout the entire city. Together with the drinking water charge, the Bad Hersfeld Clinic therefore saves €13,500 per year through the utilisation of rainwater. The operating costs, including filter maintenance and electricity for the rainwater pumps, are approximately offset by the elimination of the need to soften the cooling water. The savings in energy as a result of the installed rainwater harvesting systems also reduce CO_2 emissions, and give the hospital a smaller carbon footprint. Germany is among several industrialised countries pioneering a return of this simple but cost effective technique, while also developing rainwater capture systems in new and more sophisticated ways. By supplementing conventional supply, rainwater harvesting has the potential to reduce big, costly and sometimes environmentally-questionable infrastructure projects.

9.3. Rainwater Harvesting and Adaptation to Climate Change

The Millennium Ecosystem Assessment provided a baseline on the state of our ecosystem services. The scenario for the future looks grim given the Millennium Ecosystem Assessment's finding that 60% of the world's ecosystem services have degraded or are being used unsustainably. Many of the degraded ecosystem services were stated to be caused by increased agricultural outputs and expansion. An additional challenge is the 370 million people who were undernourished during the period 1997 to 1999. This figure increased to 852 million during the period 2000 to 2002. These marginalised groups are largely found in South East Asia and the Sub-Saharan Africa.

With the increasing rate of rise in the earth's temperature, the Fourth Assessment Report of the IPCC projected that by 2020 yields from rainfed agriculture in some countries in Africa could be reduced by up to 50%. With agricultural production and access to food adversely affected, malnutrition and hunger will increase. In Latin America, there is medium confidence in a projection that the number of people at risk of hunger will increase as productivity of important crops and livestock declines, especially in tropical and subtropical parts of the continent.

The IPCC Fourth Assessment Report further states that by mid-century, small islands can be expected to suffer from reduced water supplies to the point where they become insufficient to meet demands, especially during low rainfall periods. In Latin America, changes in precipitation patterns and the disappearance of glaciers are expected to significantly affect fresh water availability thus affecting human consumption, agriculture and energy generation. In Asia, freshwater availability is expected to decrease by 2050 affecting over a billion people. This projection covers Central, South, East and South-East Asia. The onset of water stress comes earlier in Africa. By 2020, between 75 and 250 million people are projected to be affected by increased water stress in this region, in particular in the Mediterranean region in the northern and the southern parts of the continent.

In addition to changing patterns to rainfall amounts, the rainfall events may become more intense. This may affect incidences of flooding and droughts, making the supply of freshwater ecosystem services more unreliable. The Millennium Ecosystem Assessment examined the regulating services of the ecosystem, and, out of the ten regulating ecosystem services, seven were found to be in decline. In water regulation, positive and negative impact varies depending on ecosystem change and location. However, the numbers of flood incidences continue to rise. The centralised water system is the first system to suffer collapse upon the onset of a natural disaster. The GEO4 reports that one likely impact of climate change will be higher incidence of natural disasters, such as droughts and floods. Two thirds of all natural hazards relate to hydro-meteorological events, such as floods, windstorms and high temperatures between 1992 and 2001 1.2 billion people were affected by floods. Ninety per cent of the people exposed to natural hazards reside in the developing countries.

Degradation of ecosystem services in the face of the demands of an increasing population often results in difficult trade-offs, sometimes developing into conflicts. Sometimes, too, it forces people to migrate to even more degraded areas, which further contributes to increasingly unsustainable utilisation of natural resources. This perpetuates the vicious cycle of destroying ecosystems, reducing ecosystem services for these environmental refugees, increasing their vulnerability, pushing them into abject poverty and decreasing levels of human well-being.

9.3.1. Role of Rainwater Harvesting

Rainwater harvesting is one effective water technology for adaptation to increased variability in water supply and rainfall. Its decentralised nature allows the owners to benefit from direct management of demand as well as supply. With support technologies, rainwater harvesting is cost effective, and can release capital needed in times of disasters of surprising magnitudes. There also are savings of costs related to rainwater harvesting using simple processes and therefore infrastructure, including the pumps and energy inputs needed. This also reduces greenhouse gas emissions related to water supplies. Rainwater harvesting technology can therefore contribute to both climate change mitigation and adaptation.

The Fourth Assessment Report of the IPCC itself indicated that the expanded use of rainwater harvesting and other "bottom-up" technologies have the potential of reducing emissions by around 6 Gt CO_2 equivalent/ year in 2030.

The system of water delivery in the context of current infrastructure development is part of the contributing system for green house gas emissions. The study of Flower *et al.* suggested that the main public water systems contribute to climate change by direct emissions of green house gases from water storage reservoirs and water treatment processes and through significant energy and material uses in the system. A case study from Melbourne, Australia, showed that appliances associated with residential end users of water have higher green house gas emissions than all upstream-downstream emissions.

The contribution of an urban water system to climate change comes from three sources: consumption of energy derived from carbon-based fuels, bio-diesel processes which directly generated green house gases and consumption of goods and services that involve energy consumption or biochemical generation of green house gases. The total mass of green house gas emission associated with the end user of water including upstream and downstream activities was calculated as 7,146 kg CO_2-equivalents per household per year. If we compare this carbon footprint with the most common reference made on the gas emissions of fuel used in driving a car, the car's green house gas emissions are only 4,500 kg CO_2-equivalents each year per 15,000 miles, according to the Australian Greenhouse Office. In a study of the rainwater harvesting carbon footprint in New Zealand, the green house gas emissions from the use of a rainwater tank system was estimated at 2,300 kg CO_2 equivalents per household per year. Different combinations of tanks with demand management affects the size of the CO_2 equivalent emissions related to the rainwater harvesting system used.

In Melbourne, another study was undertaken to come up with a method for achieving a climate-neutral Water Saving Framework. It was found that potable water, using rainwater and other conservation devices, generates 0.173 CO_2t/ml while a wastewater treatment plant or wastewater recycling plant generates 0.875 CO_2t/ml. Thus, saving water can save green house gas emissions. In addition, substantial green house gas savings could be made by addressing wastewater management, in addition to the rainwater harvesting intervention. In a case of a German industrial company rainwater harvesting lowered CO_2 emissions and energy and water costs. Huttinger Elektronik equipped its two-storey production and office building, built on 34,000 m^2 lot, with the following improvements:

- cooling towers for cooling work spaces
- rainwater for washing the company's street cars,
- rainwater for flushing toilets
- rain gardens through which excess rainwater percolated into the ground
- rainwater for the spray system of the re-cooling
- units of the building
- well water for cooling the offices
- rainwater for irrigation and evaporation purposes
- cool water distributed through a double-pipe system without the customary refrigeration machinery.

With the above combination of technologies, based on the use of rainwater, the company was able to save the equivalent cost of energy corresponding to 56,664 litres a year of heating oil or a reduction of an equivalent of 318 tonnes of CO_2. In the case of using rainwater for cooling purposes, the savings amounted to 98,147 litres of heating oil with an equivalent reduction in CO_2 emissions of 551 tonnes. This computation of CO_2 emission equivalents is limited to the savings generated by fuel oil used and has not included the impacts on other parts of the production system.

9.3.2. Current Adaptation Strategies

Finding No. 3 of the Millennium Ecosystem Assessment stated that the degradation of ecosystem services could grow significantly worse during the first half of this century and that such degradation is a barrier to achieving the Millennium Development Goals.

However, Finding No. 4 stated that, "*The challenge of* reversing the degradation of ecosystems while meeting increasing demands for their services can be partially met under some scenarios that the MA considered, but these involve significant changes in policies, institutions, and practices that are not currently under way. Many options exist to conserve or enhance specific ecosystem services in ways that reduce negative trade-offs or that provide positive synergies with other ecosystem services". It is also expected that many ecosystem services will be more vulnerable and fragile as climate change affects rainfall patterns and increases surface temperatures. Rainwater harvesting will continue to be one way to adapt to these increased changes in water supply and rainfall variability in the future, and, at the same time, enhance ecosystem services.

Several illustrations and case studies have been presented in the previous chapters, which highlight the contributions of rainwater harvesting to adaptation to the local challenges of water, healthy ecosystem services and human well-being. These and other cases will continue to serve as examples of adaptation strategies to climate variability, with multiple benefits. In the Philippines, farmers in rainfed areas who use rainwater collected in ponds were able to raise their production yields from an average of 2.2 tons/hectare to an average of 3.3 tons/hectare with a high of 4.68 tons/ hectare. The average yield from irrigated lands in the area is 3.3 tons/hectare. Considering the development costs of dams and irrigation canals, the government spends about US$5,000 to irrigate one hectare of land. A farmer may spend around US$400 to water one hectare of rice land. Added to the cost of the infrastructure could be the value of lost farm land, made into an irrigation pond, which the farmers estimated to be a loss equal to about 300 kilograms of rice.

Many experiences documented how rainwater harvesting conserved groundwater. The Ghogha project in rural Gujarat, India reported having successfully recharged the groundwater using 276 recharge structures in 82 villages. India maintains a Central Groundwater Board which oversees artificial recharge of groundwater both in rural and urban areas. According to Singh, the green revolution in Punjab and Haryana contributed significantly to India's food security but at the expense of soil and water degradation. The increase in groundwater use for agriculture between 1965 and 1995 resulted in a groundwater table decline of 2 meters. Alarmed by this situation, the Central Ground Water Authority planned for groundwater recharge and rooftop rainwater harvesting in another rainwater harvesting project in India. The Central Ground Water Authority reported that an additional 215 billion m3 of groundwater can be generated by harvesting and recharging only 11% of the surplus runoff.

Drinking water is another product of ecosystem services. As a result of the decline in the earth's freshwater ecosystems and the related socio-economic factors, 1.1 billion people do not have access to improved water supplies and more than 2.6 billion lack access to improved sanitation. Also, water demand has increased and water supply has decreased. The ratio of water use to accessible supply increases by 20% every ten years. Rainwater harvesting can help communities adapt to the declining availability of drinking water as droughts affect more semi-arid areas and floods inundate water sources and centralised water systems. The Thai Royal Government declared a policy of water resources development in 1979 to improve people's well-being and adapt to current rainfall variability.

The programme supported construction of jars and tanks for augmenting drinking water supplies. Because of this programme, the country was the first to attain water sufficiency during that water decade. After 10 years, 8 million tanks had been constructed. Most households have 1 ferrocement tank, a Thai jar and a membership in a community tank group. Private sector competition brought prices down and increased market availability of jars.

However, education lagged behind and incidences of diarrhea became prevalent as health measures for keeping the tanks potable were not followed. A lesson to be learned is that when business interests are strong, government control over standards and community vigilance must be monitored.

The 1981 rainwater catchment systems programme in Capiz, Philippines was accepted by local governments and the communities. The data from 2002 taken by the Planning and Development Office of the province showed an average of 67.5% the population using rainwater in 10 towns with inadequate groundwater resources. Three towns registered the 90% mark for the population using rainwater, according to the Planning and Development Office. The adoption of rainwater harvesting was a necessity to enable high quality water for domestic supply, as the groundwater was too low and sometimes too saline, and local springs were often contaminated by pollutants including agrochemicals.

The cases have shown that rainwater harvesting can make significant contributions to Millennium Development Goal No. 7 to ensure environmental sustainability. For example, the contribution of a decentralised water system such as rainwater harvesting to green house gas emission is far less than that of the current centralised water systems being used. Rainwater harvesting can contribute to water pollution control by capturing rain and using it or recharging it to groundwater. With rainwater being used, more surface water is conserved for use in aquatic ecosystem services and less groundwater is extracted. Use of rainwater harvesting for agro-forestry, in the forests, and on farms reduces soil erosion which is beneficial to the soils as well as to downstream water users.

Can rainwater harvesting mitigate soil erosion, which is one of the big drivers of ecosystems degradation? Terraces (*in situ* rainwater harvesting technologies) in the Philippines have been key rainwater harvesting technologies used especially in the upland and rainfed agricultural areas. They have helped control erosion. The recent typhoon which ravaged Panay Island and the landslides that brought down uprooted trees from the old growth forest in the mountains into the cities on the coast, showed how terraced hillsides could have withstood the landslides and excessive rains. In the experience of the Tigum-Aganan Watershed, Philippines, the flood crisis turned into a water crisis as the city's main water supply pipe from the watershed broke down. The problem continued to linger as more silt was carried by the river even 6 months after the flooding. As a result, the business of trucking water from various deep wells flourished. The urban poor suffered the most from the disrupted water supply service. Households with rainwater tanks were less affected by the water crisis. The farmers suffered too as they could not use silted irrigation water for their farms. But there were farmers who had adopted rainwater harvesting techniques who had adequate water for drinking and for growing crops during the cropping season.

Water purification is another ecosystem service rendered by the regulating function of water flowing through the landscape. Generally, the water purification function of ecosystem services is declining, according to the Millennium Ecosystem Assessment. One source of polluted water is stormwater runoff. A recent practice is to collect rainwater to prevent stormwater from bringing pollution down the drainage ways and the sewer or directly into the river and the seas. Urbanisation leads to significant changes in hydrology and pollutant transport from catchments which could harm sensitive aquatic ecosystems. An example to address the stormwater pollution in urban design is to introduce stormwater harvesting above the wetlands in Wyong-Warnervale, Australia. The design was effective in assisting the water purification process and stormwater management.

The increasing risk of natural and man-made catastrophic events has been a growing niche for water provision in a decentralised manner with limited costs. In northern Bihar, India, recurrent floods turned communities into temporary migrants, with insufficient supplies of potable water. They traditionally had abundant water supplies as a result of the multiple floods, so rainwater harvesting was not an indigenous coping strategy to access drinking water. However, when flooding occurs, wells are contaminated by floodwater. By implementing low-cost rainwater harvesting for drinking and sanitation purposes, communities and individuals stay healthy during times of crisis.

9.3.3. Effectiveness of Rainwater Harvesting

Prof. George Kuczera of the University of Newcastle studied the impact of roofwater harvesting used to supplement public water supplies in an urban setting in Sydney, Australia. He stressed the fact that water reservoirs were vulnerable to prolonged drought and climate change which reduces rainfall amounts in catchments. The consequences of such occurrences for a large urban area could be catastrophic. In Sydney, the annual rainfall average is 900 mm to 1,200 mm.

The study aimed at providing insights on the drought security performance of an integrated regional water supply and roofwater harvesting system. The base scenario, with 800 GL/year annual demand, suggested that a prolonged 10-year drought could bring about a complete failure of the system. However, with 50% of the households having a 5,000 litre rainwater tank, the probability that there will be a restriction of demand would be 8.5% to 5.2% in any year.

The probability of any household running out of water in any given year would be 0.05% to 0.02%. With a backup desalination plant, the integrated system survived the 10-year drought. The study arrived at the conclusion that roof-water harvesting can make a substantial contribution as an adaptation technology and reduce the vulnerability of water supply in urban areas.

Working from the results of earlier studies on centralised water supplies, a follow-on study proceeded to look into the efficiency of two types of water catchments in the climate change scenarios of Brisbane, Sydney, Melbourne and Perth. The result, with the exception of Perth, was that catchments exhibited a disproportionate decrease in yield in response to rainfall reductions as compared to the yields from the rainwater tanks. The authors concluded that the analysis strongly suggested that there is a significant difference in the response to climate change of the two systems for collecting water from rooftops or from catchments. The centralised water catchment systems supplying dams were seen as more sensitive, particularly in reduced runoff during reduced runoff periods, and hence potentially more susceptible to failure.

A number of cases on rainwater harvesting illustrate rainwater harvesting as an effective strategy to reduce vulnerability amongst local users to an unexpected lack of water. This suggests that one adaptation strategy to both short- and long-term variability in rainfall should be to actively work on decentralised water storage technologies, flexible to adopt and adapt in multiple user contexts. Indeed, rainwater harvesting has come of age. The technology has matured and the world condition is in such state that it needs this technology to heal itself, to protect what natural assets that have remained, and to start anew, beginning with the simplicity and appropriateness of this humble technology.

References

IPCC. *Summary for Policymakers: An Assessment of the Intergovernmental Panel for Climate Change,* Valencia Spain. 2007.

Malesu, M, Oduor, A.R., Odhiambo, O.J. eds. *Green water management handbook: rainwater harvesting for agricultural production and ecological sustainability.* Nairobi, Kenya : World Agroforestry Centre ICRAF 229p. 2008.

McGranahan G. *et al. Urban Systems. In: Ecosystems and Human Well-Being,* Current State and Trends, Findings of the Condition and Trends Working Group. 2005.

Pandey, D.N., Gupta, A.K., Anderson, D.M. "Rainwater harvesting as an adaptation to climate change". *Current Science.* Vol. 85:1 2003.

10 Rainwater Harvesting: Advantages and Pitfalls

Rainwater harvesting, in its broadest sense, is a technology used for collecting and storing rainwater for human use from rooftops, land surfaces or rock catchments using simple techniques such as jars and pots as well as engineered techniques. Rainwater harvesting has been practised for more than 4,000 years, owing to the temporal and spatial variability of rainfall. It is an important water source in many areas with significant rainfall but lacking any kind of conventional, centralised supply system. It is also a good option in areas where good quality fresh surface water or groundwater is lacking. The application of appropriate rainwater harvesting technology is important for the utilisation of rainwater as a water resource.

10.1 Need to Promote Rainwater Harvesting

In most urban areas, population is increasing rapidly and the issue of supplying adequate water to meet societal needs and to ensure equity in access to water is one of the most urgent and significant challenges faced by decision-makers. With respect to the physical alternatives to fulfil sustainable management of freshwater, there are two solutions: finding alternate or additional water resources using conventional centralised approaches; or better utilising the limited amount of water resources available in a more efficient way. To date, much attention has been given to the first option and only limited attention has been given to optimising water management systems.

Among the various alternative technologies to augment freshwater resources, rainwater harvesting and utilisation is a decentralised, environmentally sound solution, which can avoid many environmental problems often caused in conventional large-scale projects using centralised approaches.

10.1.1. Global Population Growth

Global population has more than doubled since 1950 and reached six billion in 1999. The most recent population forecasts from the United Nations indicate that, under a medium-fertility scenario, global population is likely to peak at about 8.9 billion in 2050.Given that many natural resources (such as water, soil, forests and fish stocks) are already being exploited beyond their limits in some regions, significant effort will be required to meet the needs of an additional three billion people in the next 50 years.

In parallel with these changes, there have been profound demographic shifts as people continue to migrate from rural to urban areas in search of work and new opportunities. Since 1950, the number of people living in urban areas has jumped from 750 million to more than 2.5 billion people. Currently, some 61 million people are added to cities each year through rural to urban migration, natural increase within cities, and the transformation of villages into urban areas. Urbanisation creates new needs and aspirations, as people work, live, move and socialise in different ways, and require different products and services. Urban environmental impacts and demands are also different. By 2025, the total urban population is projected to double to more than five billion, and 90 per cent of this increase is expected to occur in developing countries.

10.1.2. Global Water Crisis

Rapid population growth, combined with industrialisation, urbanisation, agricultural intensification and water-intensive lifestyles is resulting in a global water crisis. About 20 per cent of the population currently lacks access to safe drinking water, while 50 per cent lacks access to a safe sanitation system. Falling water tables are widespread and cause serious problems, both because they lead to water shortages and, in coastal areas, to salt intrusion. Both contamination of drinking water and nitrate and heavy metal pollution of rivers, lakes and reservoirs are common problems throughout the world. The world supply of freshwater cannot be increased. More and more people are becoming dependent on limited supplies of freshwater that are becoming more polluted. Water security, like food security, is becoming a major national and regional priority in many areas of the world.

10.2. Advantages of Rainwater Harvesting

Rainwater harvesting systems can provide water at or near the point where water is needed or used. The systems can be both owner and utility operated and managed. Rainwater collected using existing structures (i.e., rooftops, parking lots, playgrounds, parks, ponds, flood plains, etc.), has few negative environmental impacts compared to other technologies for water resources development. Rainwater is relatively clean and the quality is usually acceptable for many purposes with little or even no treatment. The physical and chemical properties of rainwater are usually superior to sources of groundwater that may have been subjected to contamination.

Some of the major advantages of rainwater harvesting is listed below:

a) Rainwater harvesting can co-exist with and provide a good supplement to other water sources and utility systems, thus relieving pressure on other water sources.

b) Rainwater harvesting provides a water supply buffer for use in times of emergency or breakdown of the public water supply systems, particularly during natural disasters.

c) Rainwater harvesting can reduce storm drainage load and flooding in city streets.

d) Users of rainwater are usually the owners who operate and manage the catchment system, hence, they are more likely to exercise water conservation because they know how much water is in storage and they will try to prevent the storage tank from drying up.

e) Rainwater harvesting technologies are flexible and can be built to meet almost any requirements. Construction, operation, and maintenance are not labour intensive.

10.3. Contributions to Sustainable Water Strategy

In the 1980's the world community embarked on the worthy challenge of endeavouring to provide access to clean water and sanitation to all by 1990. Even at the start of the International Drinking Water Supply and Sanitation (IDWSS) Decade (1981-1990) during which hundreds of millions of people were supplied with improved water supplies, it was clear that the goal of "water for all" was unattainable. This was due mainly to a lack of political will both in the North and the South. Another major obstacle was an unrealistic faith in modern technologies and implementation strategies which in the event proved inappropriate in many rural and peri-urban communities of the South. Despite the efforts made during and since the Decade, well over 1 billion people still lack access to convenient and safe water supplies and many more have no proper sanitation.

At the same time, another billion people have access to both abundant water and the fruits of irrigation, both of which are delivered to them at affordable and often highly subsidised prices. When the economic incentives and subsidies are available, supplying water even to desert cities such as Phoenix, Arizona is possible, although in these marginal environments such ventures are usually inherently unsustainable.

The challenge of providing water and sanitation for all is clearly one for which a solution is long overdue. Hundreds of billions of dollars have been spent on water resource development projects in developing countries in the past half century yet few of the benefits of these projects have been directed to those facing the greatest water needs. There are numerous reasons for the failure of past efforts to provide universal access to clean water including population growth, war, corruption and an insufficient commitment on the part of governments around the world to seriously address the infrastructural and social needs of either the rural or urban poor. While many new water supplies have been built, a significant number have failed due to poor management and inadequate maintenance, as well as over-exploitation, pollution or salinisation of water sources. While the introduction of new technology has improved water provision in some places, in others especially in remoter regions of developing countries, it has often proved to be unsustainable. The reasons for project failures have invariably been because the technology and/or approach used for its implementation have been technically, economically, socially, or environmentally inappropriate. At village level, past experience has shown that water systems which are dependent on external sources of fuel, spare parts, or expertise for maintenance and repair are less likely to be sustainable than those dependent only on local inputs unless well developed and reliable systems are in place for providing any external requirements.

There has been a tendency in many developing countries to equate improved water supplies with modern technologies such as large concrete dams, motorised pumps, pipeline systems which have often proved inappropriate particularly in poor rural settings. An unfortunate corollary to this has been that many traditional technologies, such as the quanat systems in Iran, tried and tested over the centuries and both sustainable and appropriate to the needs of the local community have been overlooked in favour of more modern approaches.

Rainwater harvesting systems have often been grouped into this same category of traditional technologies and have been ignored in favour of modern and supposedly better alternatives. In the Thar desert in India, many communities have depended for centuries on a variety of traditional rainwater harvesting technologies. These include tankas, simple clay lined reservoirs, kundis, covered tankas with compacted mud catchment areas and *khadins,* low walls diverting runoff from hillsides onto crops. While most of these *khadins* are still being used for runoff farming, many *tankas* and kundis have been abandoned since the arrival of modern piped water schemes. Ironically, a study by the Centre for Science and the Environment (CSE) during the 1987 drought indicated that while many on the new schemes based on tube well sources dried up, villages still depending on the traditional technologies still had water to drink.

A return to the use of traditional rainwater harvesting technologies has subsequently been actively promoted by the CSE not only in the arid states of Rajistian and Gujarat but throughout the country. Considering that world-wide there are hundreds of millions of people who depend on rainwater supplies for part or all of their domestic water needs, it is surprising how little attention the technology has been afforded. Several analysts of the growing global water crisis have also proposed that greater credence needs to be given to traditional solutions such as rainwater harvesting.

Since traditional water supply technologies have in most cases been able to meet the needs of local populations for many centuries, the systems are clearly sustainable. Ignoring this ancient wisdom and replacing traditional approaches entirely with technologies and approaches just a few decades old, which are foreign to local communities would seem unwise. Major rural water supply initiatives in both Gansu, China, and Northeast Thailand have clearly demonstrated that the traditional rainwater harvesting technologies can be upgraded and improved in order to provide affordable and sustainable supplies. These and the examples cited from Kenya and Sri Lanka support the notion that roof catchment systems could play an important role in the provision of household level water supplies in many developing countries in the coming decades. The consequences of failure to urgently provide at least the vast majority of those in rural areas with satisfactory access to clean water could be extremely far reaching. Without convenient water supplies, rural areas become even less attractive places in which to reside and the already rapid drift to the cities could accelerate putting even more pressure on the urban environment and its already over stretched water resources.

10.3.1. Roof Ccatchment Supplies

The collection of rainwater runoff from household roofs is the most common form of rainwater harvesting and attractive to householders from a several of points of view. First, for an existing dwelling the catchment area is available at no additional cost. Second, contamination of rainwater runoff from a well constructed and properly maintained roof is small compared with that from a ground catchment system. Third, roof catchments provide a water supply at the point of consumption.

Finally, since the household owns the system and is solely responsible for it maintenance is likely to be undertaken regularly, ensuring effective long-term system operation. Experience from around the world has shown that where rainwater tanks at schools, churches etc. are shared communally there operation and maintenance is often neglected. In more arid climates, household roof areas may not be sufficiently large to make investments in larger storage tanks needed for year round supply economically viable. In Kenya large surface tanks and many $90m^3$ hemi-spherical subsurface ferrocement water tanks have been constructed at hundreds of primary schools to provide potable water at schools previously adversely affected water shortages.

Inspection of a global rainfall map will reveal that in much of South and Southeast Asia, Central and West Africa, the Northeast half of South America, Central America mean annual rainfall exceeds or is close to 1 000 mm. This is significant as not only do, these regions contain over two-thirds of the planets population, but they include the regions where many of those lacking adequate water supplies currently reside.

In a semi-humid climate with a mean year round rainfall of 1 000mm, even a modest sized 50m^2 roof, can potentially yield up to 40m^3 of water annually, equivalent to more than 100 litres of water per day. Depending on the rainfall variability a tank of 2-4m^3 would probably be required to provide this level of household supply with a reasonable level of reliability. While in a semi-arid climate, with seasonal mean rainfall of 500mm/a, even from a modest sized 50m^2 roof, potentially 20m^3 of water could be collected which with sufficient storage could supply more than 50 litres of water per day to the household. This would, however, require a storage capacity of perhaps 5-10m^3 depending on the degree of rainfall variability to ensure a high reliability of supply. While the costs of roofwater harvesting increases in drier, more seasonal climates due to the larger catchment area and storage volume requirements, so do the costs of other alternatives.

In climates subject to drought it is not uncommon to find that governments are forced to deliver emergency water supplies by truck if boreholes, reservoirs, or other water sources dry up. Trucking water can be extremely expensive and in situations where such supplies are used from time to time, there may often be a justification for installing rainwater tanks on economic grounds alone. Rainwater can be stored for long periods of time provided that light, insects, animals, and organic matter are excluded from the tank.

In more humid climates, especially where year round rainfall is available, a large proportion of household water demand can normally be met with only a relatively small storage tank. This was clearly demonstrated by recent analysis of the performance of household roof catchment systems in two villages in Uganda. Mbarara and Kyenjoro are subject to two rainy seasons and receive mean annual rainfall totals of 900mm and 1400mm, respectively. The analysis revealed that in both cases a storage capacity equivalent to just a four days of household consumption could provide a useful domestic water source for at least half of the year. In the light of these findings it is easy to explain the enormous popularity shown for the relatively small (1-2m^3) rainwater jars in Northeast Thailand where the rainfall conditions and large roof areas make rainwater harvesting even more favourable. Clearly, in regions where rainfall conditions are conducive to roofwater harvesting relatively small investments in storage tanks can produce significant impacts.

10.3.2. Rainwater harvesting in rural areas

The failure of both traditional and improved communal systems to provide adequate water supplies in rural areas is an important reason why household rainwater collection is gaining popularity in many developing countries e.g. China, Thailand, Kenya, and Sri Lanka. Probably the single most important factor, however, is the shift away from the use of traditional roof construction techniques involving grass/palm thatch and dried mud in favour of modern impervious roofing materials such as fired clay tiles and corrugated iron. This improvement has resulted in households possessing appropriate catchment surfaces which only require a tank and simple gutter in order to provide a rainwater supply. The development of appropriate and affordable designs such as those for the in situ concrete and ferrocement tanks in several countries has further encouraged the spread of rainwater systems.

The benefits of rainwater collection in the rural context of the developing world is particularly significant for women due to the great amount of both time and energy which they can save. In many parts of rural Africa and Asia women still spend hours each day collecting water. In some cases and especially in dry periods this will involve treks to polluted sources several kilometres from people's homes. Under such circumstances the acquisition of a household rainwater tank represents an immediate and dramatic improvement in quality of life. For example, even a modest $2m^3$ Thai rainwater jar can store the equivalent of one hundred 20 litre water buckets weighing 2 000kg (two metric tons) which without the tank the householders, usually the women and children, would have had to collect and carry some distance. Given this reality, even if rainwater supplies cannot meet the total household requirements any water they can provide represents a substantial benefit to the household.

Increasingly, rural women are collecting improved water supplies from communal tap stands now found in most larger and many smaller villages in the South. Nevertheless, even where these are available women may still have to walk several hundred metres or more to collect water and sometimes queue for long periods to fill each bucket. Breakdowns and the drying up of supplies in drought periods are also common occurrences in some regions. Given such circumstances, the introduction of rainwater tanks may be appropriate even when communal taps are available.

The provision of water at the point of consumption from rainwater tanks provides a range of immediate positive social impacts on health, family welfare and domestic productivity. This results when time saved in water collection is utilised elsewhere. Some of the time saved maybe used for productive activities such as agriculture with clearly tangible and easily valued economic benefits. More time can also be spent on activities such as child rearing when women have time freed up from the daily chore of water collection. The value of such benefits to family livelihood and well-being are difficult to assess and are rarely appropriately costed.

The use of rainwater harvesting for domestic water supplies is also relevant in the rural areas of many more developed countries e.g. Australia, New Zealand and the USA.. In Australia rainwater use has always been practised by farming households in the outback. Currently, over 1 million people still depend on rainwater for all or part of their domestic water needs. Formal guidelines with advice to householders on best practice regarding rainwater tank use were recently produced.

These guidelines include advice on protecting water quality, water treatment, maintenance and repair and tank sizing including tables to assist with calculations. In South Australia where rainwater collection is widely practised even in urban areas such as Adelaide and in much of the arid interior of the continent and especially Western Australia large mechanically graded surface catchments often many hectares in area are used to collect water for livestock and small settlement supplies. Rainwater harvesting is widely used around the world for purposes other than domestic water supply. Rainwater runoff is often be diverted onto plots or into ditches or ponds to provide water for gardening, tree nurseries, aquaculture, livestock watering and numerous other purposes.

10.3.3. Rainwater Harvesting in Urban Areas

Traditionally most rainwater harvesting has been most commonly associated with remote rural areas or those lacking alternative water sources, e.g. coral islands. Several global trends are making the more general use of rainwater utilisation in future more likely especially in large cities. The most significant of these trends is urbanisation especially in developing countries. There has already been an almost doubling of the urban population in the South since 1960. By 2030 it is expected that there will be nearly 5 billion people living in urban areas world-wide of which more than 75% will be in Asia, Africa, and Latin America.

Urban development is already putting considerable strains on existing water resources which are struggling to keep pace with steadily rising demands. The continuing over-exploitation and pollution of many existing water sources is leading to a growing interest in alternatives such as rainwater catchment systems as supplementary water sources with multi-purpose functions.

Japan and Germany are both leading the way with respect to developing models for future patterns of rainwater utilisation in cities. Flooding and problems associated with ground subsidence related to groundwater extraction are currently faced by several large Asian megacities including Tokyo, Osaka, Shanghai and Bangkok and are fuelling this interest in rainwater utilisation. In Germany there are even deliberate attempts to encourage householders to collect rainwater runoff and divert any unneeded surplus to recharge groundwater. The utilisation of rainwater in an urban context provides several potentially beneficial functions with respect to the following:

- *Flood control*—by greatly reducing urban runoff;
- *Stormwater drainage*—by reducing the size and scale of infrastructure requirements;
- *Firefighting and disaster relief* —by providing independent household reservoirs;
- *Water conservation*—as less water is required from other sources;
- *Reduced groundwater exploitation and subsidence*—as less groundwater is required;
- *Financial savings*—where rainwater can be used in place of water purchased from water vendors often charging up to 10-50 times the official water tariff.

Potentially perhaps the most significant impact of rainwater harvesting technology in the urban context may be in providing water supplies to hundreds of millions on unserved residents on the peripheries of the new megacities of the South such as Tegucigalpa in Honduras.

Predictions regarding Global warming could have a major effect in significantly increasing water demand in many cities. At the same time increased evaporation from reservoirs and reduced river flows in some areas may decrease the available surface water supplies. A greater uncertainty regarding yields from major reservoirs and well fields is likely to make investments in the diversification of water sources, better water management and water conservation even more prudent in future. Increased climatic variability and the greater frequencies of droughts and floods possible in many areas will also make the role of rainwater harvesting systems even more important as sources of supplementary, back-up, or emergency water supply. This will particularly be the case in areas where increasing pressure is put on existing water resources.

Many large cities around the world already face periodic water shortages and millions of urban residents including those connected to reticulated systems are forced to buy water from vendors or retailers and often to ration water supplies. As cities continue to grow in the future such problems are likely to become increasingly common. Since cities comprise of numerous impervious surfaces designed to encourage rainwater runoff the scope for rainwater collection is substantial.

Atmospheric pollution remains a major constraint as it contaminates both the rainwater and catchment surfaces making rainwater unsuitable for drinking in many cities around the world. Nevertheless, rainwater can still be used for non-potable uses such as toilet flushing, clothes washing and gardening. Furthermore, great use of rainwater in urban areas could in future significantly strengthen the lobby to clean up the urban atmosphere entirely.

10.4. Demand management

Although strictly speaking a rainwater harvesting technology itself is a supply side option which can be used to help reduce pressure on other conventional water sources. In many respects the operation and management of a roof water system can act as an important demonstration of the renewable yet finite nature of water resources and the need to use water in a rational and sparing way. Rainwater catchment systems can therefore often provide a central focus for any integrated water conservation campaign in which both demand and supply side approaches are adopted. The public's level of awareness regarding the importance of using water resources sparingly and rationing usage in times of drought will be greatly enhanced if they have first-hand experience of managing their own rainwater supply.

10.4.1. Community Control/Involvement

Just as community participation and control is essential to the successful implementation, operation and maintenance of any rainwater tank project, it is equally important with respect to any effort to encourage demand management. Ultimately demand management strategies are unlikely to succeed without strict individual self-discipline and community control / policing with respect to any directives or preferably recommendations agreed by the community themselves regarding efforts to promote water conservation.

10.4.3. Public-private Initiatives

There is some evidence that combined public-private sector approaches for rainwater harvesting initiatives can work effectively in certain circumstances, e.g. the Thai jar programme. The role of the private sector for funding programmes in the poorest regions such as Africa and South Asia where most of the unserved communities reside should nevertheless be viewed with some caution. Despite much rhetoric at the World Bank about the role of the private sector in community water supply and sanitation, very little of its investment sector is directed to the poorest communities or the poorest regions, notably Africa and South Asia.

10.4.4. Technological Advances

To gain the full benefit of a rainwater supply which is limited by the available rainfall, catchment area, and storage capacity it is generally appropriate to use the system in conjunction with water saving technologies and practices. This is especially the case in arid environments where water scarcity will be greatest particularly during the dry season and in drought periods. Technological advances in tank construction techniques, such as the development of cheaper ferrocement water tank technology and especially the Thai jar have greatly assisted the rate at which rainwater catchment systems have been adopted. Future innovations particularly with respect to the further development of low cost durable water tank designs could encourage an even faster spread of the technology.

10.4.5. Raising Public Awareness

Public awareness and education with respect to the importance of water conservation and the potential role which rainwater harvesting can play is still at a very basic level in many parts of the world. A significant proportion of householders who harvest rainwater could greatly improve the performance of their systems through very small and inexpensive design changes such as ensuring that runoff from the whole of the roof catchment area is being diverted towards the tank and not just part of it.

10.5. Managing Rainwater Harvesting Projects

With respect to household roof catchment systems the issue of affordability is a crucial one especially within the context of the rural areas of the developing world. The cost of a rainwater tank could typically be equivalent to the household's total cash income for a year and such large outright payments are beyond the means of most households. This presents a major barrier preventing individuals from installing roof catchment systems independently especially in more drier seasonal climates where greater storage capacities are needed. For households in poorer communities, some form of financing mechanism preferably in combination with an appropriate subsidy will often be the only possible way of acquiring a rainwater system.

Examples of some successful approaches to project financing such as the use of revolving funds in Kenya and the provisions of subsidies in Germany are detailed in Boxes 3 and 6, respectively. Other approaches which have been used successfully elsewhere include linking household rainwater harvesting initiatives to income generating projects, e.g. Philippines and Indonesia, where pairs of young breeding goats and piglets have been included in a rainwater tank construction financing package. If the animals breed successfully, two of the young are returned to the donor, some are retained for breeding and some fattened up for sale to generate income for any loan repayment. While this approach contains some risks if breeding is unsuccessful, when it works well it can help to make projects virtually self financing.

Individual household rainwater systems often seem expensive compared to development of communal rural shallow well or protected spring water supplies. This is generally the case and if good quality shallow groundwater or spring water are available rainwater collection may only be appropriate as a supplementary source. In the early 1980's much effort was directed into developing increasingly low cost designs and while some of these were successful, others such as the bamboo reinforced cement as a low-cost alternative to ferrocement failed to stand the test of time.

Although there are some notable exceptions, the development and spread of rainwater technology has not generally been strongly promoted by government water departments or most international donors whose major funding has remained focused on conventional and large-scale developments, e.g. dam construction and reticulated supplies. In terms of meeting the challenge of providing water supplies to the billion plus people still unserved, this is significant as large dams are seldom built to meet the needs of either the rural or urban poor. Furthermore, the communities most affected by major dam and other water resource developments are rarely, if ever, fully engaged in project planning and frequently barely consulted at all. In some cases, even where rainwater projects have been supported by governments or donors, the communities affected have been little involved. This invariably leads to project failures such as occurred in Kilifi, Kenya in the mid-1980's.

Even when rainwater utilisation is compared on an apparently equal basis with all other feasible water supply options, it is usually subject to an inadvertent negative bias. This is because the standard economic analysis used for determining the costs and benefits of major dam developments usually under-estimate the full cost. For example, many externalities such as the natural ecological services provided by a river prior to damming are seldom fully costed and sometimes ignored altogether. Another common omission is the cost of the eventual decommissioning of the dam. The use of inappropriate market based rather than social based discount rates can also distort the true cost of a project.

In many situations the issue will not be one of whether rainwater harvesting should be considered as a substitute for another water supply option, but how could its use as a supplementary water source be integrated into a project to develop the most cost effective overall mix of technologies for maximising the benefits and reliability of the supplies.

Despite the high initial cost of developing roofwater supplies in rural areas, when compared with the level of spending on some major urban water supply developments even in relatively poorer regions such as Africa the costs seem modest. In Botswana, a recently completed 350km long North-South Carrier pipeline and new Letsibogo dam constructed almost exclusively for municipal water supply for the capital and a few other urban centres with a total population of less than 400,000 has cost around $250 million. This is more than $500 per capita. A similar level of investment in the construction of household rainwater harvesting systems could probably have provided domestic water supplies albeit of lower quality and quantity to ten times as many people in scattered rural areas. To take an even more extreme example one might consider weighing the benefits of spending $5 billion dollars on constructing a

Man-made River Project in the Libyan Desert for irrigation with using the funds to construct household roof tank systems and simple runoff farming plots and microcatchments for 20 million rural African families.

10.6. Institutions and Processes for Assessing Options

In most countries, the potential use of rainwater harvesting along with a broad range of both supply side and demand side technological and management options available for helping to ensure that future water needs are met are still given at best only cursory consideration. In this respect the current processes for assessing water supply options should be considered to be seriously inadequate.

The use of roofwater, stormwater, and several other potentially significant alternative water sources such as the reuse of greywater and water conservation strategies including the installation of water saving technologies which could have a significant impact on curbing water demand are still frequently not even considered during standard water planning exercises in many countries. There seems no rational reason that rainwater harvesting and a range of less conventional options should not warrant consideration for providing either a supplementary or total water supply along with all other available options.

In many instances even where the exclusive use of rainwater tanks may not be economic, the technology may prove to be a cost effective alternative during critical drought periods when conventional sources dry up or when there is a system breakdown. In a few countries, the implications of developing water resources sustainability is beginning to lead to gradual changes in thinking and approach in some institutions although due to inertia and pressures from commercial interests and consumers this is a slow process.

Clearly there is a need for training and capacity building in most institutions dealing with water resource development to encourage broader, long-term and enlightened thinking to ensure the implementation of technologies and approaches which will address the fundamental challenge to the sector faces in the new century of providing adequate and sustainable water supplies to all.

Practitioners within existing government water ministries and departments as well as planners and decision makers need to be encouraged to think laterally and consider a wider range of potential solutions to water provision including less conventional approaches such as rainwater harvesting. If the predicted future problems of water scarcity are to be minimised a new approach to water resource planning is needed in which the starting point should be the starting point in any master plan should be the size of the available water resource that can be supplied at an economic and sustainable level. A range of approaches including demand and supply side technology and management options, should then be considered adopted to see how best the demand and target supply can be tailored together.

Slowly the implications of developing water resources within the new framework of sustainable development is leading changes and their is a gradual recognition around the world of the possibilities for approaching the task in radically different way. This enlightened thinking still seems to be limited to a minority of institutions and individuals but seems to be spreading.

10.7. Issues in Rainwater Harvesting

One of the most important underlying values in rainwater harvesting is that it is a benign technology and cannot create undesirable consequences. Water harvesting initiatives are driven by firm beliefs and assumptions, some of which are: 1) that there is a huge amount of monsoon flow, which remains un-captured and eventually ends up in the natural sinks, especially seas and oceans, supported by the national level aggregates of macro hydrology; 2) that local water needs are too small and as such exogenous water is not needed; 3) that local water harvesting systems are always small and, therefore, are cost-effective; 4) since the economic, social and environmental values of water are very high in regions hit by water shortages, water harvesting interventions are viable, supported by the assumption that cost- effective alternatives that can bring in the same amount of water, do not exist; 5) incremental structures lead to incremental benefits; and 6) being small with low water storage and diversion capacities, they do not pose negative consequences for downstream uses.

10.7.1. Lack of Emphasis on Local Water Demand

Rainwater harvesting ignores a few critical parameters that govern the potential of rainwater harvesting systems (RWHS) in meeting local water demand, such as: a) the hydrological regime of the region/locality; b) the reliability of the supplies, governed by the reliability of rainfall; c) the constraint imposed by local geological and geo-hydrological settings on recharge potential; and d) the aggregate demand for water from various sectors within the local area. Some basic hydrological phenomena, which make the above mentioned parameters very critical in deciding the scope of rainwater harvesting and groundwater recharging, are:

— For runoff harvesting, rainfall has to exceed a threshold to generate runoff, though the threshold would vary according to the nature of the soil and land cover of the area. The estimated runoff based on the regression equation arrived at from observed flows in the Hathmati subbasin of the Sabarmati Basin ($R=0.00193*X^{2.022}$) in western India, shows that for the runoff to cross 100 mm, the minimum rainfall required is 682 mm. Whereas in the case of the Kabani subbasin of the Cauvery Basin, runoff starts when the rainfall crosses 366 mm. However, the actual runoff rates would depend on how strong is the correlation between rainfall and runoff in a given basin, and this relation weakens if there is a major year to year change in rainfall intensity and pattern.

- Regions with lower mean annual rainfall experience higher variability and vise versa. Hence, in regions with lower mean annual rainfalls, rainwater harvesting as a dependable source of water is likely to be low.
- Generally, it has been found that a greater magnitude of annual rainfall means a larger number of rainy days and smaller magnitude of annual rainfall means fewer number of rainy days spread over the rainy season. The examples of Gujarat further illustrate this. Fewer rainy days also means longer dry spells and thus greater losses from evaporation for the same region.
- High intensity rainfalls are common in the semi-arid and arid regions of India. Higher intensity of rainfall can lead to high intensity in runoff, occurring in short durations, limiting the effective storage capacity of rainwater harvesting systems to almost equal their actual storage size.
- High evaporation during the rainy season means losses from surface storage structures. It also means a faster rate of soil moisture depletion through both evaporation from barren soils and evapotranspiration, which increase the rate and quantum of soil infiltration. This reduces the generation potential of runoff.
- Soil infiltration capacity can be a limiting factor for recharge. In sandy and sandy loam soils, the infiltration capacity of the recharge area can be sustained through the continuous removal of soils. But clayey soils have inherent limitations.
- For artificial recharge, the storage potential of the aquifer is extremely important. The storage potential of an aquifer vis-à-vis the additional recharge is determined by the characteristics in geological formations, and the likely depth of the dewatered zone.
- In hilly watersheds, the area available for cultivation is generally very low, keeping agricultural water demand low. At the same time, the surface water potential available for harvesting is generally high due to high

rainfall and runoff coefficients. On the contrary, towards the valleys and plains, the area available for cultivation increases, raising agricultural water demand. At the same time, the surface water potential available for harnessing is generally low due to the lower rainfall, and low runoff coefficients owing to mild slopes, high PET and deeper soil profiles.

The implications of some of these factors on the potential of rainwater harvesting systems are analysed in the following two sections.

10.7.2. Limitations Imposed by Hydrological Regimes

Local water management interventions are often based on very little understanding of the local hydrological regimes, which govern the potential supplies of water for harvesting. They are rather based on the deep-rooted belief that the greater the size of the water impounding structure, the greater would be the hydrological benefit in terms of water storage and recharge. The best example is the participatory water conservation movement launched by the Government of Gujarat. The government implemented large-scale work for the excavation of thousands of village ponds, irrespective of the nature and size of catchments. Part of the reason is the lack of availability of data on inflows, determined by stream-flows; and outflows, determined by evaporation rates, for small rainwater catchments. While runoff harvesting is most suited to areas with a high 'runoff catchment area' to 'run on' area ratio, this is also ignored. The higher the aridity, the larger would be the required catchment area to the cropped area required for the same water yield. Often, encroachment of catchments of water harvesting systems for crop cultivation is very rampant, reducing the runoff prospects.

10.7.3. Limitations Imposed by the Socioeconomic System

Water harvesting arguments totally miss the water demand-availability perspective at the micro level. Ideally, the RWHS would work if the area which has uncommitted flows to harness has an 'un-met demand' or vise versa. This is unlike in large water resource systems where provisions exist for the transfer of water from 'surplus' areas to deficit areas.

The water demand of an area is determined by the agro-climate and existing socioeconomic system, which, in fact, gets adjusted by the natural resource environment of the village, the available technologies for accessing them and the institutional and policy environments over a period of time. Regions that were heavily into irrigated agriculture in the past, supported by good water endowments, institutional support and favourable policies, might continue demanding large quantities of water for irrigation even when they run out of water. This is because communities take quite some time to devise coping and adaptive strategies to manage with conditions of water deficits.

Studies in a village in Mandvi taluka of Kachchh, which is one of the most arid districts in India, showed that the annual water withdrawal from aquifers for irrigating crops is 25.42 MCM. The entire water requirement in the village was being met by groundwater, which is experiencing severe over-draft conditions. The total amount of rainwater falling in the village is nearly 10.14 MCM. With a surface water potential of 0.014 MCM/sq. km, the amount of runoff water that would be available for replenishment through natural and artificial recharge from within the village is only 0.40 MCM. The runoff is, therefore, a small fraction of the total consumptive use. This means that the village has to depend on exogenous sources of water for making water use sustainable. What is presented is representative of almost the entire peninsular of India excluding Kerala, central India and western India.

In a village named Manund, in the Patan District of North Gujarat, which has seen widespread pond de-silting, the total groundwater abstraction for agriculture alone was estimated to be 3.78 MCM (or 275 mm), with 35 deep tubewells pumping water at a rate of nearly 15,000 gallons per hour for nearly 1,500 hours a year.

The groundwater condition of the village is typical of the North Gujarat region. Against this, the total amount of rainfall over the village is only 7.56 MCM, with a mean annual rainfall of 550 mm over an area of 1,374 ha. The runoff that this amount of rainfall can generate is 63.8 mm as per the rainfall runoff relationship, with the total runoff being 0.877 MCM. But in practice, it is unlikely to get this amount of runoff, as farmers directly harness a significant portion of the runoff generated from the crop land, which falls in the catchment 'in situ' for crop production, unlike large basins which have a good part under virgin catchments. Kumar estimated the groundwater overdraft in the village as nearly 247.5 mm by considering the recharge as 5 % of the annual rainfall. Hence, even if the entire runoff generated is harnessed for recharge, it would amount to only 25.7 % of the overdraft.

On the other hand, there are many regions in India where the economic demand for water is far below what the natural endowment can provide. The entire Ganga-Brahmaputra Basin area can be put into this category. The region has an enormous amount of static groundwater, estimated to be 8,787.6 BCM, apart from having a high rainfall and cold subhumid climate that generate sufficient surface flows. Cheaper access to water might increase the demand for irrigation water slightly, but, there are significant limits to such access, imposed by the cold and humid climate and very low per capita arable land. The economic demand for water therefore, would continue to be below what the water endowment can provide.

Already, the irrigation intensities are high in thee Uttar Pradesh and Haryana. Though irrigation intensity in Bihar is low, the subhumid and cold climate reduces the irrigation requirement significantly. In most parts of this region, the issue is not the physical availability of water, but the ability of communities to access water for irrigation.

Water harvesting anyway does not offer any economic solution here for the poorer communities to facilitate their access to water.

10.7.4. Issues in Evaluating Costs and Economics

In the planning of large water resource systems, cost and economics are important considerations in evaluating different options. But unfortunately, the same does not seem to be applicable in the case of small systems, though concerns about economics of recharge systems in certain situations were raised by authors such as Phadtare and Kumar.

Part of the reason for the lack of emphasis on 'cost' is the lack of scientific understanding of the hydrological aspects of small-scale interventions, such as the amount of stream flows that are available at the point of impoundment, their patterns, the amount that could be impounded or recharged and the influence area of the recharge system. Even though simulation models are available for analysing catchment hydrology, there are great difficulties in generating vital data at the micro level, especially those on daily rainfall, soil infiltration rates, catchment slopes, land cover and PET, which determine the potential inflows; and evaporation rates that determine the potential outflows. Furthermore, for small water harvesting projects, implemented by local agencies and NGOs with small budgets, the cost of hydrological investigations and planning is hard to justify. Often, provision for such items is not made in small water harvesting projects.

That said, the amount of runoff that a water harvesting structure could capture, depends not only on the total quantum of runoff, but also on how it occurs. A total annual runoff of 20 cm occurring over a catchment of one sq. km can generate a surface flow of 0.20 MCM, but the amount that could be captured for water harvesting depends on the pattern of rainfall events. As Garg points out, in arid and semi-arid regions in India, high-intensity rainfalls of short duration are quite common. These runoffs generate flash floods. If the entire runoff occurs in a major rainfall event, the runoff collection efficiency would reduce with the reducing capacity of the structures built. If large structures are built to capture the high-intensity runoff and thereby increase the runoff collection efficiency, the cost per unit volume of water captured would inflate. In fact, authors such as Oweis, Hachum and Kijne have argued that runoff harvesting should be encouraged in arid areas only if the harvested water is directly diverted to the crops for use.

Scale considerations are extremely important in evaluating the cost and economics of water harvesting/ groundwater recharge structures because of the hydrological integration of catchments at watershed and river basin levels. The cost and economics of water harvesting systems cannot be performed for individual systems in isolation, when the amount of surplus water available in a basin is limited. This is because incremental structures do not result in a proportional increase in hydrological benefits, as interventions in the upper catchments reduce the potential hydrological benefits from the lower systems. What is important is the incremental hydrological benefits due to the new structure. A system in itself may be cost-effective and economically viable if evaluated independently, but, if evaluated as a part of a large-scale water-harvesting intervention at the level of river basins, the system may not be justifiable from the cost angle when compared against the additional benefit it brings in.

In any basin, the marginal benefit from a new water harvesting structure would be smaller at higher degrees of basin development, while the marginal cost would be higher. The reason being: 1) the higher the degree of basin development is, the lower would be the chances for getting socially and economically viable sites for building water impounding structures, increasing the economic and financial cost of harvesting every unit of water; and 2) with a higher degree of development, the social and environmental costs of harvesting every unit of water increases, reducing the net economic value of benefits. Therefore, the cost and economic evaluation should move from watershed to basin level.

There are regions where human and cattle drinking become high priority demands. North western Rajasthan, which is arid and dominated by pastoral communities, named Gujjars, is one such example. The social and economic value realised from the use of water for human drinking and livestock use, respectively, would be much more than the economic value realised from the use of water in irrigating crops. In such situations, water should be diverted for such uses where the opportunity costs are low and net value products are high. But proper water use planning to realise the maximum value from the added water is largely missing in water harvesting efforts.

10.7.5. Lack of an Integrated Approach

In many river basins, the surface water systems and groundwater systems are often interconnected. Any alterations made in one type of system could change the availability of water in the other type. In many hilly areas, especially in the Western Ghats, the water levels rise steeply after the monsoon, and groundwater contributes significantly to the stream flows downstream during lean seasons due to the steep gradients for groundwater flow. In such cases, any water harvesting intervention to store water underground may not make much sense as the water stored would be rejected and appear as surface flows. On the other hand, in regions with deep water table conditions like in North Gujarat, the runoff directly moves into the groundwater systems of the plains through the sandy river bed as dewatering of the upper aquifers increases the rate and quantity of percolation.

With two-third of the country's geographical area underlain by hard-rock formations, the storage capacity of aquifers poses a major challenge for artificial recharge. Most parts of water-scarce states, viz., Gujarat, Madhya Pradesh, Maharashtra, Karnataka, Andhra Pradesh, Orissa, Chhattisgarh and Tamil Nadu are underlain by hard rocks ranging from basalt, crystalline granite, hill aquifers and sandstone. A small area in Gujarat has extensive alluvium e.g., Narmada Valley and Cambay Basin.

The hard rock aquifers have no primary porosity and have only secondary porosity. The constraints imposed by hardrock geology in recharge efforts through percolation tanks are: high depth to the water table below and around the recharge structure due to the occurrence of recharge mounts and shallow bed rocks, which prevent the percolation of water; and low infiltration capacity of the thin soils overlaying the hard-rock formations. Due to low specific yield (0.01-0.03), the sharp rise in water levels is observed in aquifers during monsoon, leaving little space for infiltration from structures. While harnessing water for recharge is extremely important during normal and wet years, the natural recharge in hard-rock formation is high during such years as it is a function of seasonal rainfall, further reducing the scope for artificial recharge.

In Saurashtra, in spite of the poor potential offered by low rainfalls, high variability, and high evaporation rates, significant recharge efforts were made. But, the biggest constraint in storing water underground during high rainfall years is the poor storage capacity or specific yield of the basalt formations. During good rainfall years, the aquifers get saturated with natural recharge immediately after the rains, leaving no space for entry of water from the recharge systems.

An estimated 20,000 check dams built in the region to capture the rainwater and recharge the aquifers are able to store only a small fraction of the surplus runoff. In such situations, proper water use programming is required to achieve effective utilisation of the available surplus water, wherein water from aquifers is pumped out and used during the rainy season itself thereby creating storage space for the incoming flows.

The groundwater level fluctuation data obtained from the Ghelo River basin in Saurashtra illustrate this. The basin had experienced intensive water-harvesting since 1995. The data were collected from open wells located inside the basin periodically during and after the monsoon rains. The wells located close to the water harvesting structures and those away from the structures are demarcated.

10.7.6. Maximising Benefits

Generally, in any river basin, the upper catchments are rich in terms of their ability to contribute to the basin yields. This is mainly because of the unique physiographical features, and partly because of climatic conditions such as steep slopes, high rainfall in the mountains and high humidity, which provide a favourable environment for runoff generation. The upper catchments also provide a good source of base flows due to the forest cover, which causes favourable conditions for water storage and infiltration.

On the demand side, these regions generally are less endowed in terms of the availability of arable land. Over and above, the demand rates for irrigation are generally low. On the other hand, the lower catchments are generally characterised by lower rainfalls and higher levels of aridity and better access to arable land, thereby increasing the aggregate demand for irrigation.

There are numerous examples for this and a few to cite are: the upper catchment of the Cauvery Basin in the south, the Narmada Basin in central India, the Sabarmati Basin in western India, the tributaries of the Indus River in north western India, the Krishna Basin in central India and the Mahanadi Basin in eastern India. Certain parts of the Kabani Subbasin of the Cauvery River basin have a cold and semi-humid climate, while certain other parts of this subbasin receive the second highest rainfall in India after Chirapunji, with the mean annual rainfall exceeding 4,000 mm. Irrigation demands in these regions are low owing to high precipitation and low reference evapotranspiration, and the low per capita availability of arable land. On the other hand, the lower parts of the Cauvery Basin in Tamil Nadu are hit by a scarcity of water for irrigation owing to lower rainfalls and high evapotranspiration. We have defined agricultural water demand as a function of per capita net sown area and the ratio of ET_0 and rainfall; and water availability as a function of rainfall. It is assumed that: a) higher the ET_0/R ratio, higher would be the irrigation requirement for a unit of land; and b) higher the per capita net sown area, higher would be the aggregate demand for irrigation per capita.

Major water resource/irrigation projects undertaken in the past, tap stream flows generated from the upper catchments but cater to either the lower parts of these basins or other less water endowed regions outside these basins. The Bakhra Reservoir and Nangal diversion projects located in the high rainfall Shivalik Hills of Himachal Pradesh, essentially cater to the ravenous low rainfall and drought-prone regions of Punjab and scanty rainfall regions of Rajasthan. The Sardar Sarovar Dam harnesses water from ample rainfall areas in the Narmada Valley and takes it to the drought-prone areas of North Gujarat and Saurashtra, which are characterised by low and erratic rainfall.

Similarly, the large reservoir projects in the Cauvery Basin transfer water to the drought-prone regions in Tamil Nadu and Karnataka. As such the water demand for irrigation is extremely low in the upper catchments. Moreover, as irrigation water use efficiency and water productivity are likely to be high in areas with variability in rainfall and high drought- proneness, with transfer of water from the well-endowed regions to the poorly-endowed regions, the economic value of water in agriculture increases. The recent research carried out by IWMI in water-scarce and land-rich western Punjab and water-rich and land-scarce eastern Uttar Pradesh (UP) showed that the value of water realised from irrigation is much higher in Punjab than in eastern UP.

The economic value of water was Rs. 14.85/m^3 in western Punjab, whereas in eastern Uttar Pradesh it was Rs. 11/m^3. Due to the scarcity of water, the farmers in Punjab make better economic use of water by choosing cropping systems that are economically more efficient and adopting agronomic practices in order to obtain higher yields, higher physical productivity and economic efficiency.

But, often water harvesting initiatives, especially those by NGOs, are driven by considerations other than economic efficiency, the most important of which are social equity and environmental justice. Impounding water in the upper catchments might serve the social objectives of meeting drinking water requirements.

As evident from the above illustrations, there is a clear trade off between meeting economic efficiency objectives and these developmental goals. Therefore, any water resource intervention in the upper catchment areas that reduces the downstream uses should be done with due consideration to the net change in the 'gross value product' of water in the basin due to the interventions.

The 'gross value product' can be defined as the sum total of the incremental value product from the economic uses, environmental services and social uses that the basin's water resources meet. The amount of water to be captured upstream through RWH interventions should also be optimised to derive maximum regional social equity, environmental value and overall output from the economic uses of water. In basins where the available water resources are already committed, the challenge is bigger as maximising the gross value product might mean reallocating some water from a low valued use to a high valued use.

10.8. Promoting Rainwater Utilisation

Rainwater utilisation, together with water conservation and wastewater reclamation, should be incorporated into municipal ordinances and regulations. Some standardisation of materials, at least at regional level, may be desirable from a maintenance and replacement point of view. It may be also appropriate to standardise the design of the rainwater utilisation system, at least at the regional level.

Various implementation policies should be established to make rainwater utilisation and other measures a part of the social system. Leadership is very important and local governments must take the initiative to promote the concept of water resource independence and restoration of the natural hydrological cycle. Consideration should be given to subsidising facilities for rainwater utilisation.

Encouraging technology and human resources development to support rainwater utilisation is very important. It is also important to promote the development of efficient and affordable devices to conserve water, facilities to use rainwater and devices to enhance the underground seepage of rainwater. Together with this, there is a need to train specialists with a thorough grasp of these technologies and devices.

To promote rainwater harvesting and utilisation as an environmentally sound approach for sustainable urban water management, a network should be established involving government administrators, citizens, architects, plumbers and representatives of equipment manufacturers. It is essential to encourage regional exchanges amongst public servants, citizens and industry representatives involved in rainwater storage, seepage and use, as well as the conservation and reclamation of water

References

Agarwal, A.; Narain, S. *Dying Wisdom: Rise and Fall of Traditional Water Harvesting Systems,* Centre for Science and Environment. New Delhi, India: Centre for Science and Environment. 1997.

Kumar, M. D. *Food Security and Sustainable Agriculture in India: The Water Management Challenge,* Working Paper 60, International Water Management Institute, Colombo, Sri Lanka. 2003.

Muralidharan, D.; Athawale, R.N. Artificial Recharge in India, base paper prepared for Rajiv Gandhi National Drinking Water Mission, Ministry of Rural Areas and Development, National Geophysical Research Institute. 1998.

Varghese, B. G. Water and Energy for Future India, paper presented at the International Regional Symposium on Water for Human Survival, 26-29, November, 2002, Hotel Taj Residency, New Delhi. 2002.

11 Policy Environment for Rainwater Harvesting in India

In India, rainwater harvesting has been in practice for more than 4000 years owing to the temporal and spatial variability of rainfall, in its broadest sense, a technique used for collecting and storing rainwater for human use from roof tops, land surfaces or rock catchments. Simply stated a water harvesting system collects and stores water within accessible distance of its place of use. The physical and chemical properties of rainwater are usually superior to the sources of ground water that may have been subjected to contamination. Like other water resources, rainwater harvesting is an option to be considered when planning a community oriented water supply system. Depending on local environmental conditions, water harvesting may provide a supplementary supply, an alternative supply or the only feasible improved supply, especially in urban areas.

The current centralised water supply paradigm seems unsustainable and extremely high on energy consumption. As an alternative paradigm for more sustainable water availability harvesting rainwater, storing it in tanks, and recharging groundwater may be put in place. The first step could be to initiate domestic rooftop rainwater harvesting. In a house the highest point is the rooftop. However, common property resource issues have to be considered at larger level for rainwater harvesting. An additional benefit is that the experience in this technique is considerable. The government has issued policy directives for encouraging this practice.

On the civil society becoming more aware and sensitised regarding its potential, rainwater harvesting can perhaps be scaled up to neighbourhood and micro-watershed levels.

Armed with complete information, it has been seen that people naturally would have a tendency to conserve their own resources. Therefore, a review of the approaches to rooftop rainwater harvesting as a practice of local water management, with a focus on the findings (and the failures) appears to be relevant for putting policy into practice.

11.1. Policy Framework and Implementation

Managing freshwater scarcity constitutes to be one of the biggest responsibilities of governance everywhere in the world, and thus, local management of this resource is an indispensable component. Many countries have national water laws. In India, however there are no national laws as such, though there is a national water policy. The National Water policy 2002 is a cogent and comprehensive document and forms a basis that could be converted into a law.

In India, under the constitutional set up, water is a state subject. In urban areas its governance rests with urban local bodies in their areas of jurisdiction as per the 74th constitutional amendment The need for a policy framework for water harvesting systems arises mainly because the prevailing policy statements do not touch extensively upon the issue. There is a clear need to evolve a decentralised legal regime with regard to water, which empowers people and makes them real managers of resources. For promoting urban water harvesting, a policy should include a mix of incentives and penalties. Measures that need to be undertaken include:

- Rainwater harvesting/recharge of ground water system should be an essential town planning requirement and a pre-requisite for permission for the development of new colonies.
- Provision of rainwater and harvesting structures in all building plans should be mandatory for issuing of building permission.
- Appropriate rebates on property/fiscal incentives should be granted for effective implementation of rainwater harvesting systems.

A number of state governments have made rainwater harvesting compulsory for new buildings according to their plot sizes in various Indian cities.

11.1.1. Policy of Government of Madhya Pradesh

The Government of Madhya Pradesh also formulated the State Water Policy in 2003. The Departments of Housing & Environment and Urban Administration and Development, Government of M.P. have issued various notifications and directives to urban local bodies for ensuring rainwater harvesting. These are described below:

Roof top rainwater harvesting has been enforced in municipal areas through the Madhya Pradesh Bhumi Vikas Rules, 1984. Rule 78 (4) inserted vide order no. F-23(107)-95-XXXII, dated 07.04.2000 of the Housing and Environment Department, which provides that for rooftop rainwater water harvesting, methods, shall have to be provided on all type of buildings having plot size more than 500 sq m. This has been amended vide order F-23(107)-95-XXXII (1), dated 30.05.2001 of the Housing and Environment Department reducing the minimum plot size for mandatory rooftop rainwater harvesting from 500 sq m to 250 sq m. The Article 138 of M.P. Municipal Corporation Act, 1956 and the Article 126 of M.P. Municipalities Act, 1961 has been amended by order F-5-5-2001-XVIII (3), dated 13.03.2001 of Urban Administration and Development Department providing a rebate of 6 % to the property tax for the year in which the construction of rainwater harvesting facility has been completed.

11.1.1.1. Rainwater Harvesting Measures Prevailing in Municipal Corporations

11.1.1.1.1. Municipal Corporation Gwalior

By order dated 27th January 2006 rainwater harvesting has been made mandatory for buildings having an area more than 250 sq m. The engineer in charge of the area has been authorised to impose a penalty of Rs. 7000 in case of noncompliance. A rebate of 6 % in property tax in the year in which the construction of rainwater harvesting facility has been completed will be provided for the owner of the building as an incentive.

11.1.1.1.2. Municipal Corporation Bhopal

The Corporation has not taken any action to date.

11.1.1.1.3. Municipal Corporation Indore

Rainwater harvesting has been made mandatory in all new buildings with an area of 250 sq m or more from 2001-02. A rebate of 6 per cent on property tax has been offered as an incentive for implementing rainwater-harvesting systems in the building

11.1.1.1.4. Municipal Corporation Jabalpur

Rainwater harvesting has been made mandatory in all new buildings with an area of 250 sq m or more from 1 April 2005. A rebate of 6 per cent on property tax hạs been offered as an incentive for implementing rainwater-harvesting systems in the building.

11.1.1.2. Status of the Measures

From the analysis of present status, it is clear that the implementation is lacking and none of the cities have taken any strong steps to ensure compliance of the order issued by the government. Despite the provision of the rebate, only 522 building owners have availed of the rebate which indicates the low order of this incentive. On examining the cost of construction of rainwater harvesting structures versus the financial incentive being provided one comes the scenario as below:

The cost of rainwater harvesting structures depends on the type of technology adopted. The normal cost is around Rs 3000. This cost consists of the cost of pipes, filter media and construction of a pit. Out of this, pipes have to be laid for draining the roof water. Hence the additional cost for the construction of Rrainwater harvesting structure would be Rs 1500 including the cost of the filter media and construction of the pit. Coming to the incentives for an average plot size of 250 sq m having two storied building, the property tax comes out to Rs 5000 per annum. The rebate of 6% of the amount will be of Rs 300 only. Therefore the present subsidy is about one fifth of the incremental cost of the rainwater harvesting structure and is not enough to motivate the public.

Possible reasons for the lack of success of the measures on the field are:

a) No penal provision for not constructing rainwater harvesting structures,

b) The Municipal Corporations not exercising its authority on colonisers, engineers and architects for ensuring rainwater harvesting measures in the development of colonies, design; construction & supervision of buildings.

c) Lack of sustained and mass based campaign for generation of awareness among the public.

d) Lack of demonstration centres for advising the general public in the technicalities of construction of rainwater harvesting structures.

11.2. Sociopolitical Support for Rainwater Harvesting

Rainwater Harvesting (RWH) is one of several components of water resources management. It is based on the principle that as rain is decentralised, its harvesting need also be decentralised RWH is dependent on two factors – extent and periodicity of rainfall, and the surface on which it falls. The latter includes natural surfaces such as land mass, vegetation, water bodies on land, seas, etc. and man made structures. To the extent that rainfall and land surfaces vary from country to country, region to region, the need for and extent of rainwater harvesting should also vary. Varied also is the importance given to RWH as a technique to access fresh water.

In India, the nature of rainfall is completely different from that of middle latitude countries of Europe, where precipitation occurs in the form of rain and snow throughout the year, providing a year round underground seepage. This comparison is important because the dominant technologies and administrative and political systems that are associated with water resources management in India today are largely a colonial legacy.

The rainfall pattern of India was not understood by the colonial engineers from Europe, and 'Indian engineers continue to make the same mistakes and build projects that are inappropriate to the native environment'. Unlike a number of European countries, India does not receive rainfall all through the year. Pisharoty, a leading Indian meteorologist, points out the following characteristics of rainfall in India.

- It is largely concentrated during four months of the year.
- Here too, rainfall does not occur daily. In most parts of the country, there is precipitation during not more than 50 days.
- On the days when rainfall occurs, rainfall does not fall over a period of 24 hours. In fact, heavy showers are common.
- Most of the country receives rain for just about 100 hours each year.
- The thumb rule is that the number of hours of rain a place receives is equal to the number of centimeters of rain it receives annually.

The significant features of the monsoon rainfall patterns are that most of the country receives rain for about 100 hours each year, half of which is received in only 20 hours. Since very little precipitation in India is in the form of snow, harvesting rainwater when it falls and where it falls is critical to sustain all life–supporting systems and activities.

In essence, RWH is indispensable to India as the first priority and starting point in water resources management;

it requires an integrated approach to land and water management; it requires a decentralised approach. Questions arise as to whether and how far this fundamental reality was understood and acknowledged by societies in India in different historical periods and the nature of policy and legal arrangements that they adopted for RWH and the success of such arrangements.

Pre-modern societies of India were very conscious of the necessity of RWH. Traditional water technologies in India offer a proof of this awareness. Archeological evidence in the form of excavated physical structures as well as currently functional water systems that are tens of hundreds of years old, testify to a variety of technologies that were developed and successfully utilised all over the country. These technologies also prove that harvesting of rainwater through surface or underground storages required an integrated land and water management approach that was suited to the highly varied climatic, topographical, geological and ecological characteristics of the different regions in India.

The initiators and promoters of water technologies ranged from rulers to citizens, evidencing commonly exercised access rights. The numerical predominance of small-scale water technology in pre-colonial India points to the practice of decentralised water management that implied the exercise of a wide range of local rights and powers of control over water resources. While large-scale artificial lake systems, tank systems, canal systems, river diversionary systems, etc. were constructed by medieval rulers, they were overshadowed in number and spread by systems created by individuals, local communities, or local bodies.

The jurisprudential-legal system in ancient India also supported RWH, decentralisation of land and water resource control and management. It is best understood in comparison with modern jurisprudence. The modern or formal legal framework relating to natural resources in India has its origin in the colonial period. The jurisprudential framework underlying formal natural resources law is the Anglo-Saxon legal system, the foundation of the Indian legal system today. Under this system, the three identified sources of law are *statute, precedent* and *doctrine* in that order of precedence, with statute bearing over-riding authority over the other sources. Wherever precedent or doctrine contradicts statute, however old the former, the latter will prevail.

The modern *formal* legal framework is thus entirely statute-centred. This is of particular importance in the realm of natural resources management in India, where there prevails a significant level of combination of old and new technologies and institutions. The older technologies and institutions have their origin in a much earlier and entirely different jurisprudential base. The governance of natural resources, which was essentially decentralised in character, had its legal basis almost entirely in *Custom*.

In traditional Indian jurisprudence, custom constituted a source of law *independent* from all other known sources i.e. ethical doctrine or *dharma,* or royal decrees. In pre-modern Indian society, the undisputed law - making and judicial authority of the King did not permit him to modify either *dharmic* laws or common customs of localities and groups. The King was prohibited from interfering with the customs of countries, castes and families and other groups. In addition to this prohibition, the State was also expected to 'act' in support of customary law. Extensive autonomy in the functioning of guilds, corporations, artisans, religious fraternities, local governing bodies etc. characterised the structure of the State. The conventions of these bodies were recorded in writing, constituting their statutes or by-laws.

The period of colonial rule saw a transformation of policy and law on natural resources. This transformation relates to local property rights in land and water, local authority in control and management of resources, the role of the State in collecting agricultural surplus and investing in local production systems, the political and social relationship between village and state – changes in all of which had a direct and debilitating impact on RWH. An irreversible pattern of exploitative land, water and forest use was put into place, the fundamental features of which have been retained substantially till date.

11.3. Modern Policy Approaches

Environmentally sustainable development as a fundamental aspect of and special approach to *political* governance was not envisaged in India at the time of creation of the Constitution of India, which is the fountainhead of natural resource law in the country. As a corollary, natural resources statutes – land, water and forest legislations – of colonial origin, founded on the colonial policy of resource *exploitation,* were legitimised by the Constitution and continued into the post-Constitutional era in a substantially identical form. It was not until 1976 that concept of environmental protection and development was first given Constitutional recognition by a declaration of the duty of citizens and state to protect the environment and wildlife resources of the country. However, it was too little too late.

The conceptual approach to RWH in current national and state policy therefore remains embedded in colonial perspectives of India's natural resources profile and their administration. This is well reflected in the National Water Policy, 2002. While the National Water Policy cannot be prescriptive by reason of water resources being exclusively under the legislative competence of the States, as a guideline, it fails completely to shift from the old "project oriented" approach to water resources. It is a policy for the consumption and utilisation of the State apparatus and not for the people at large. It also fails to provide a shift in perspective towards understanding the dependence of the country on the monsoonal pattern of rainfall and what should be the appropriate response to it nationally.

For instance, the NWP states that "precipitation is confined to only about three or four months in a year and varies from 100 mm in the western parts of Rajasthan to over 10000 mm at Cherrapunji in Meghalaya" It fails to carry the logic further and acknowledge the narrower interval of time in which the rain actually falls, measured in terms of hours. If this were to be acknowledged, it would entail an obligation to acknowledge the necessity of a vast effort across the country in "monsoon harvesting" within a short window of time, no different from time-bound wheat harvesting or paddy harvesting. Such an effort would require large scale devolution of responsibilities and authority to communities across the country which is anathema to the "year round project oriented" approach of Government agencies. Typically, the section on Participatory Approach to Water Resources Management envisages people's participation, not in the governance of water resources, but "in various aspects of planning, design, development and management of the water resources *schemes*."

That MH is not central to water resources management policy at the national level, is evident from the fact that it is mentioned in several contexts as an addendum to other priority approaches. For instance it is stressed mostly in relation to "drought prone areas", without acknowledging its need everywhere. It is mentioned as one of the "traditional methods", next in priority to other non-conventional methods. In other words, RWH remains at the level of piecemeal strategy, rather than an objective of water resources management.

The NWP ignores the observations of another Government body – the Steering Committee on Irrigation for the Tenth Five Year Plan (2002-2007), Government of India Planning Commission –whose report of May 2002 states "Most of the rainfall 76 percent of it as per India Meteorological Department - in India occurs as a result of the southwest monsoon between June and September, except in the State of Tamil Nadu which falls under the influence of the northeast monsoon during October and November. More than 50 percent of precipitation takes place in about 15 days and less than 100 hours altogether in a year"

Watershed management has been mentioned in the NWP in relation to water resources planning, not as something which is indispensable, but one which should be "promoted". This particular approach enables undertaking watershed development projects in a piecemeal way in selected locations across the country, rather than adopting it as a fundamental approach to natural resource management. A matter of related significance is that "land use policy" has been mentioned only once in the NWP, under the section on "Irrigation". The need for integration of land use policy with water resources policy as a whole has been lost sight of. The concept of "Water Zoning "has been included, without reference to land use zoning.

Since water resources is a "State subject" under the Constitution, a brief review of water law and policy at the State level would be useful in assessing the importance of MH Water policies at the State level reflect the same sectoral, project oriented approach to water resources management, which constitutes a barrier to RWH. A few state policies are reviewed. The Madhya Pradesh Water Policy talks of RWH only in relation to "water scarcity areas"; watershed development and water zoning is discussed only in the context of Irrigation; the technologies referred to in the policy are only "dams and canals"; traditional water harvesting structures are termed "minor irrigation tanks" and mentioned only in the context of recharging groundwater, ignoring their other functions.

The Karnataka Water Policy devotes a single line to RWH, quoting the NWP: "Rainwater harvesting and water conservation will be encouraged". It is mentioned in connection with promoting conservation consciousness. There is a total absence of perspective of incorporating RWH in all aspects of water resources planning and implementation. The Uttar Pradesh Water Policy includes a brief reference to RWH – "Water harvesting should be given consideration in planning water resources. Viable projects, specially in scarce ground water areas, should be investigated and implemented to increase the surface water availability would also help in recharging the ground water." From this it is evident that RWH is seen not as an approach to WRM or NRM, but as specific projects to be implemented in select locations.

The Rajasthan Water Policy projects a slightly better perspective than the others referred to above. It contains a section on "Maximising Water Availability" and lists several strategies to achieve this. One of them is "traditional water harvesting practices shall be preserved and encouraged" and the other: "Projects for artificial recharge of ground water shall be prepared". However, there is a need to expand the concept of RWH beyond "traditional water harvesting" and mainstream it in all aspects of developmental planning. For instance, the section on "Water Conservation and Efficiency of Utilisation" which discussed the topic in different sectoral contexts-domestic sector, agriculture, industry, watershed development, *does not once mention RWH*. None of the state policies reviewed contain objectives and strategies for urban water resources management as distinct from rural.

With respect to legislation on water and other natural resources, almost all States have on their statute books natural resources statutes such as Land Reform laws, Land Revenue Codes or Acts, Land Improvement Acts, Soil Conservation Acts, Land Development Acts, Forest Act, Irrigation Acts, legislation for Drinking Water, etc.

None of the natural resources statutes have the avowed objective of managing natural resources sustainably. They are mostly designed for the optimum exploitation of resources. Regulatory provisions in these statutes are linked to the same purpose of resource exploitation. Property laws enable inheritance or transfer of resource rights. However, such rights have not been made subject to the sustainable use of the resource. Norms of sustainability of resource use may only be defined at a local level, on scientific assessments of resource availability and demand. The antiquated resource legislations that are in force do not provide for any such management models. Legal rights therefore are totally abstracted from sustainable management practices.

State level land, water and forest legislation establish independent, mutually exclusive administrative frameworks for each of the resources separately contradictory to the interests of integrated management of resources, which is required for MH. These statutes are mostly 'command and control' laws, with almost no devolution of powers to enable decentralised management of resources. Some recent efforts at "participatory management" transfer responsibilities without any authority. The powers of allocating legal rights including recognising customary rights to natural resources is exercised monopolistically by State authorities, far removed from local contexts of management and conservation Local customary laws and practices in resource 'management' are completely unacknowledged. These statutes do not provide for a broad-based institutional framework that offers scope for a spectrum of stakeholders to participate in the governance of natural resources. Programme strategies devised by external agencies/non-government organisations in partnership with government, which introduce participatory institutional mechanisms, almost always do so on the strength of *executive orders* rather than *statutory reform*, which are rarely sustained beyond the project period. More importantly, the role of the State is not modified in any significant way. This legal regime prevents the emergence and sustenance of local legal frameworks that can support more decentralised governance of natural resources which is required for MH.

Powers of control of all *land* resources are vested in the State, with statutory authorities vested with all powers necessary for administration of land resources. Participatory institutional frameworks for the administration of land resources do not feature in any of the land laws. District – level panchayat institutions have, in some states been conferred with administrative powers over land resources. However this has been done within the rubric of existing land laws, which do not reflect an 'integrated natural resource management' perspective, and which retain the residual powers of the State Governments over the resource.

Significantly, with the exception of Kerala, no State in the country has a land use policy. In most States, natural resource laws, by conferring monopoly legal powers on government to collect and utilise revenue from land, water and forest resources have deprived local bodies of the important power of taxation.

A review of ground water legislation in the country – the "model" national groundwater bill as well as statutes in selected states shows that ground water law is entirely focused on the regulation of wells, and have no relation whatsoever with land and water resources management in general. These laws too, declare the State's exclusive powers of regulation of wells, which is practically impossible to exercise.

In spite of the Constitutional Amendments on Panchayat Raj for local self governance, in none of the states has planning and regulatory powers over *natural resources per se,* been devolved on local bodies. They are involved only planning and implementation of State Govt. or Central Government – determined schemes and programmes which are in any case contradictory to integrated natural resource management.

Rainwater harvesting in urban areas has been accorded policy and legal support in but a handful of states. Changes in municipal laws with state wide application have been made only in Kerala and Tamil Nadu; specific urban bodies in other states have been covered by legal provisions – Mumbai, Delhi, Indore; Hyderabad. In all these States, the focus has been on mandating RWH in mostly private and in some places public *buildings,* through incentives and disincentives, ignoring the vast urban land resources as a whole. These efforts are unlinked to policy and law on urban land use in totality, or even in most cases with urban water supply and drainage policy and law.

A new and emerging threat to participatory, decentralised and sustainable management of natural resources, of which MH is a critical component, is the

new buildings with a roof area of more than 100 sq m and in all plots with an area of more than 1000 sq m, that are being developed. The Central Ground Water Authority (CGWA) has made rainwater harvesting mandatory in all institutions and residential colonies in notified areas (South and southwest Delhi and adjoining areas like Faridabad, Gurgaon and Ghaziabad). This is also applicable to all the buildings in notified areas that have tubewells. The deadline for this was for March 31, 2002. The CGWA has also banned drilling of tubewells in notified areas. Draft Delhi water board (amendment) Bill.

11.4.8. Indore (Madhya Pradesh)

Rainwater harvesting has been made mandatory in all new buildings with an area of 250 sq m or more. A rebate of 6 per cent on property tax has been offered as an incentive for implementing rainwater harvesting systems.

11.4.9. Kanpur (Uttar Pradesh)

Rainwater harvesting has been made mandatory in all new buildings with an area of 1000 sq m or more.

11.4.10. Hyderabad (Andhra Pradesh)

Rainwater harvesting has been made mandatory in all new buildings with an area of 300 sq m or more. Tentative for enforcing this deadline was June 2001.

11.4.11. Tamil Nadu

Through an ordinance titled Tamilnadu Muncipal Laws ordinance, 2003, dated July 19, 2003, the government of Tamil Nadu has made rainwater harvesting mandatory for all the buildings, both public and private, in the state. The deadline to construct rainwater harvesting structures is August 31, 2003.

11.4.12. Haryana

Haryana Urban Development Authority (HUDA) has made rainwater harvesting mandatory in all new buildings irrespective of roof area. In the notified areas in Gurgaon town and the adjoining industrial areas all the institutions and residential colonies have been asked to adopt water harvesting by the CGWA. This is also applicable to all the buildings in notified areas having a tubewell, deadline was for March 31, 2002. The CGWA has also banned drilling of tubewells in notified areas.

11.4.13. Rajasthan

The state government has made rainwater harvesting mandatory for all public and establishments and all properties in plots covering more than 500 sq m in urban areas.

11.4.14. Mumbai

The state government has made rainwater harvesting mandatory for all buildings that are being constructed on plots that are more than 1,000 sq m in size. The deadline set for this was October, 2002.

11.4.15. Gujarat

The state roads and buildings department has made rainwater harvesting mandatory for all government buildings.

References

Athawale, R. N. *Water Harvesting and Sustainable Supply in India,* published for Environment and Development Series, Centre for Environment Education, Rawat Publications, Jaipur and New Delhi. 2003.

Government of India. *Integrated Water Resource Development: A Plan for Development,* Report of the National Commission for Integrated Water Resources Development Vol. I, Ministry of Water Resources, Government of India, New Delhi. 1999.

Pisharoty, P. R. *Characteristics of Indian Rainfall,* Monograph, Physical Research Laboratories, Ahmedabad. 1990.

Rangachari, R.; Sengupta, N.; Iyer, R.R.; Banergy, P.; Singh, S. *Large Dams: India's Experience,* Final Report, prepared for the World Commission on Dams, Cape Town, South Africa, Secretariat of World Commission on Dams. 2000.

Bibliography

Agarwal A.,and S. Narain. *Dying wisdom: Rise, fall and potential of India's traditional water harvesting systems* 4th edition. Eds., State of Indias Environment, a citizens' report 4, Centre for Science and Environment, New Delhi, 2005

Arya, S.L., Samra, J.S. 2001. *Revisiting watershed management institutions in Haryana Shivaliks, India.* Central soil and Water Conservation Research and Training Institute, Chandigarh, India, pp. 329.

Athawale, R. N. *Water Harvesting and Sustainable Supply in India,* published for Environment and Development Series, Centre for Environment Education, Rawat Publications, Jaipur and New Delhi. 2003.

Badiger, S.; Sakthivadivel, R.; Aloysius, N.; Sally, H. Preliminary Assessment of a Traditional Approach to Rainwater Harvesting and Artificial Recharging of Groundwater in Alwar District, Rajasthan, paper presented at the annual partners' meet of IWMI-Tata Water Policy Research Program, Anand, February 19-20. 2002.

Barron, J., Noel, S. et al. *Agricultural water management in smallholder farming systems: the value of soft components in meso-scale interventions.* SEI Project Report, Stockholm Environment Institute, Stockholm. 2008

Boers, Th M, and Ben-Asher, J. "A review of rainwater harvesting". *In Agric. Water Management* 5:145-158. 1982.

Evenari, M, Shana, L and Tadmor, N H. *The Neger, the challenges of a desert.* Havard University Press. Cambridge, Mass. 1971.

Frasier, Gary, and Lloyd Myers. *Handbook of Water Harvesting.* Washington D.C.: U.S. Dept. of Agriculture, Agricultural Research Service, 1983.

Geerts, S., Raes, D. "Deficit irrigation as an on-farm strategy to maximize crop water productivity in dry areas". *Agric. Water Manage* 96, 1275-1284. 2009.

GEO4. Global *Environmental Outlook 4: Environment for development.* United Nations Environment Programme, Nairobi/ Progress Press, Malta. 2007

Gould, J., and E. Nissen-Petersen. *Rainwater Catchment Systems for Domestic Supply: Design, Construction and Implementation.* London, U.K.: IT Publications, 1999.

Government of India. *Integrated Water Resource Development: A Plan for Development,* Report of the National Commission for Integrated Water Resources Development Vol. I, Ministry of Water Resources, Government of India, New Delhi. 1999.

Hadwen, P. "Caribbean Islands: A Review of Roof and Purpose Built Catchments." In *Non-Conventional Water Resources Use in Developing Countries.* New York, United Nations, pp. 455-468. (Natural Resources/Water Series No. 22). 1987.

Hemenway, Toby. *Gaia's Garden: A Guide to Home-Scale Permaculture.* Vermont: Chelsea Green Publishing Company, 2000.

Hudson, N. Soil and Water Conservation in semi-arid areas. FAO Soils Bulletin No. 57. FAO, Rome. 172 pp. 1987.

IPCC. *Summary for Policymakers: An Assessment of the Intergovernmental Panel for Climate Change,* Valencia Spain. 2007.

Kumar, M. D. *Food Security and Sustainable Agriculture in India: The Water Management Challenge,* Working Paper 60, International Water Management Institute, Colombo, Sri Lanka. 2003.

Lowes, P. "The Water Decade: Half Time". in in John Pickford (ed.). *Developing World Water.* London: Grosvenor Press International. pp. pp 16-17. ISBN 0-946027-29-3. 1987.

Ludwig, Art. *Create an Oasis With Greywater: Choosing, Building, and Using Greywater Systems.* California: Oasis Design, 1994.

Lye D. "Health risks associated with consumption of untreated water from household roof catchment systems". *Journal of the American Water Resources Association* 38(5):1301-1306. 2002.

Lye D. 2002. "Health risks associated with consumption of untreated water from household roof catchment systems". *Journal of the American Water Resources Association* 38(5):1301-1306.

Macomber P. "Guidelines on rainwater catchment systems for Hawaii". College of Tropical Agriculture and Human Resources, University of Hawaii at Manoa. 2001.

Malesu, M, Oduor, A.R., Odhiambo, O.J. eds. *Green water management handbook: rainwater harvesting for agricultural production and ecological sustainability.* Nairobi, Kenya : World Agroforestry Centre ICRAF 229p. 2008.

McGranahan G. *et al. Urban Systems. In: Ecosystems and Human Well-Being,* Current State and Trends, Findings of the Condition and Trends Working Group. 2005.

Millennium Ecosystems Assessment (MA). *Ecosystems and human well-being: synthesis.* Island Press, Washington D.C. 2005.

Muralidharan, D.; Athawale, R.N. Artificial Recharge in India, base paper prepared for Rajiv Gandhi National Drinking Water Mission, Ministry of Rural Areas and Development, National Geophysical Research Institute. 1998.

Pacey A., and A. Cullis. *Rainwater Harvesting: The Collection of Rainfall and Runoff in Rural Areas.* London, U.K.: IT Publications, 1986.

Pandey, D.N., Gupta, A.K., Anderson, D.M. "Rainwater harvesting as an adaptation to climate change". *Current Science.* Vol. 85:1 2003.

Pisharoty, P. R. *Characteristics of Indian Rainfall,* Monograph, Physical Research Laboratories, Ahmedabad. 1990.

Proceedings of the First International Conference on Rainwater Cistern Systems. Honolulu, Hawaii, University of Hawaii, Water Resources Research Center. 1982.

Raju, K. C. B. Rainwater Harvesting for Groundwater Recharging: The Experience of SVRTI in Moti Rayan Village, Kachchh in Moench, M. (Ed.) *Groundwater Management: Supply Dominated Focus of Government and NGO Efforts,* Monograph, VIKSAT, Nehru Foundation for Development, Ahmedabad. 1995.

Rangachari, R.; Sengupta, N.; Iyer, R.R.; Banergy, P.; Singh, S. *Large Dams: India's Experience,* Final Report, prepared for the World Commission on Dams, Cape Town, South Africa, Secretariat of World Commission on Dams. 2000.

Sahuquillo, A. Groundwater in Water Resource Planning: Conjunctive Use, *Water International,* 10. 57-63. 1985.

Sharma, B.R, et.al., (eds.). *Watershed Management Challenges: Improving productivity, resources and livelihoods.* International Water Management Institute, Colombo, Sri Lanka. 2005.

Sharma, P, *Ferrocement Bhandaran Patra (Ferrocement Storage Tanks)—National Drinking Water Mission,* SERC, Ghaziabad, 1990.

SWMRG, Evaluation and Promotion of Rainwater harvesting in Semi-Arid areas of Tanzania Research Project. 2nd interim Technical report. Sokoine University of Agriculture, Morogoro. 80 pp. 1995.

Texas Water Development Board. *Texas guide to rainwater harvesting.* Austin (TX): Texas Water Development Board. 58 p. 1997.

Thomas PR, Grenne GR. "Rainwater quality from different roof catchments". *Water Science Technology* (28):290-99. 1993.

Thomas PR, Grenne GR. 1993. "Rainwater quality from different roof catchments". *Water Science Technology* (28):290-99. EPA, www.epa.gov/safewater/mcl.html

Thomas, E.N. "The Artificial and Roof Rainwater Catches of Bermuda." Paper presented at the UN/CSC Seminar on Small Islands Water Problems. Barbados. 1980.

UNEP. *Rain and Stormwater Harvesting in Rural Areas.* Nairobi, pp. 92-113. (Water Resources Series, 5) 1983.

United Nations. Department of Technical Cooperation and Development. *Roof Catchments, Roof Coverings, Guttering and Downpipes.* New York. 1989.

Varghese, B. G. Water and Energy for Future India, paper presented at the International Regional Symposium on Water for Human Survival, 26-29, November, 2002, Hotel Taj Residency, New Delhi. 2002.

Zaunderer, J and Hutchinson, C F. A review of water harvesting techniques of the Arid Southwestern US and North Mexico. Working paper for the World Bank's Sub-Sahara Water Harvesting Study. 1988.

Index